HEINEMANN MATHEMATICS 4

Textbook

These are the different types of pages and symbols used in this book and associated workbooks.

I
Numbers to 99: number names

Most textbook and workbook pages are of this type. They deal with mathematical concepts, skills and applications in number, measure, shape and handling data.

6
Other activity: number puzzles

These pages provide self-contained activities which need not necessarily be tackled in the order in which they are presented. They are intended to give further opportunities for children to apply the mathematics they have learned or to extend their experience.

Problem solving

Some pages, or parts of a page, provide an opportunity for problem solving or investigative work.

Extension

Some pages, or parts of a page, contain work which either extends the range of a mathematical topic for many children or provides more difficult examples suited to more able pupils.

Where a calculator would be useful this is indicated by a calculator symbol.

Use hundreds tens and units if you wish

This shows that structured materials may be used.

HEINEMANN EDUCATIONAL

Contents

Heinemann Educational,
a Division of Heinemann Publishers (Oxford) Ltd,
Halley Court, Jordan Hill, Oxford OX2 8EJ

© Scottish Primary Mathematics Group 1993

First Published 1993   ISBN 0 435 02154 0

Photographs by Chris Coggins. Our thanks to Botley
County Primary School, Oxford, for their co-operation.
Designed by Miller, Craig and Cocking
Produced by Oxprint
Printed in the UK by Bath Colour Books, Glasgow

93   94   95   96   97   98   10   9   8   7   6   5   4   3   2

MYSTERY MANOR

73 OAK STREET

Jenny, Ali, Paul and Ruth visit Mystery Manor. It is at number seventy-three Oak Street.

1 Write these other house numbers in words.

65 Hawthorn Hall

~21~ Minton Mansion

88 TORWOOD TOWERS

56 Greystone Grove

2 Name the house which has
- (a) the tens digit five
- (b) the units digit five
- (c) the same tens and units digit
- (d) the smallest tens digit.

3 Draw number signs for these houses.
- (a) Corrie Court – number forty-nine
- (b) Grindlay Grange – number seventeen
- (c) Cannon Castle – number thirty-two
- (d) Huntly House – number ninety-four

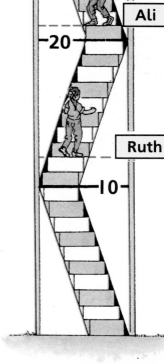

Jenny has climbed 26 steps of the taller tower.
26 is 30 **to the nearest ten**.

Ali has climbed 21 steps of the shorter tower.
21 is 20 **to the nearest ten**.

1 **(a)** How many steps has Paul climbed?
  (b) Write this number to the nearest ten.

2 **(a)** How many steps has Ruth climbed?
  (b) Write this number to the nearest ten.

3 Write each number to the nearest ten.

(a) 53 **(b)** 48 **(c)** 64 **(d)** 19 **(e)** 86
(f) 32 **(g)** 77 **(h)** 91 **(i)** 26 **(j)** 13

4 **(a)** How many steps altogether are in the taller tower?
  (b) Write this number to the nearest ten.

5 **(a)** How many steps altogether are in the shorter tower?
  (b) Write this number to the nearest ten.

The taller tower has 39 steps.
The shorter tower has 32 steps.

> 39 is about 40. 32 is about 30.
> 40 + 30 = 70, so altogether there
> are **about** 70 steps in the 2 towers.

1 Use Ali's method to do these.

(a) 42 + 37   (b) 26 + 53   (c) 74 + 16   (d) 28 + 14
(e) 27 + 49   (f) 11 + 63   (g) 61 + 18   (h) 33 + 59

> 39 is about 40. 32 is about 30.
> 40 − 30 = 10, so there are
> **about** 10 more steps in the taller
> tower than in the shorter tower.

2 Use Jenny's method to do these.

(a) 52 − 29   (b) 38 − 11   (c) 61 − 28   (d) 89 − 13
(e) 72 − 14   (f) 41 − 27   (g) 77 − 16   (h) 98 − 22

3 In the garden, there are 41 elm trees and 17 oak trees.

(a) **About** how many trees are there altogether?
(b) **About** how many more elm trees are there
    than oak trees?

Paul adds 39 and 32.

He writes:

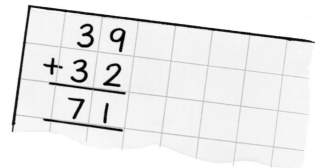

$$
\begin{array}{r}
3\,9 \\
+\,3\,2 \\
\hline
7\,1
\end{array}
$$

Altogether there are **exactly** 71 steps in the 2 towers.

1 **(a)**
$$
\begin{array}{r}
4\,2 \\
+\,2\,7 \\
\hline
 \\
\end{array}
$$
 (b)
$$
\begin{array}{r}
3\,9 \\
+\,4\,3 \\
\hline
 \\
\end{array}
$$
 (c)
$$
\begin{array}{r}
5\,1 \\
+\,1\,8 \\
\hline
 \\
\end{array}
$$
 (d)
$$
\begin{array}{r}
2\,6 \\
+\,1\,7 \\
\hline
 \\
\end{array}
$$
 (e)
$$
\begin{array}{r}
1\,5 \\
+\,7\,5 \\
\hline
 \\
\end{array}
$$

Ruth subtracts 32 from 39.

She writes:

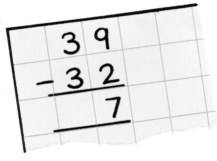

$$
\begin{array}{r}
3\,9 \\
-\,3\,2 \\
\hline
7
\end{array}
$$

There are **exactly** 7 more steps in the taller tower.

2 **(a)**
$$
\begin{array}{r}
7\,9 \\
-\,4\,5 \\
\hline
 \\
\end{array}
$$
 (b)
$$
\begin{array}{r}
3\,6 \\
-\,2\,0 \\
\hline
 \\
\end{array}
$$
 (c)
$$
\begin{array}{r}
9\,9 \\
-\,6\,9 \\
\hline
 \\
\end{array}
$$
 (d)
$$
\begin{array}{r}
6\,2 \\
-\,9 \\
\hline
 \\
\end{array}
$$
 (e)
$$
\begin{array}{r}
4\,0 \\
-\,2\,6 \\
\hline
 \\
\end{array}
$$

3 **(a)** 64 + 26   **(b)** 43 − 8   **(c)** 66 + 6   **(d)** 90 − 27

4 In the garden there are 54 rose bushes and 29 holly bushes.

(a) How many bushes are there altogether?

(b) How many more rose bushes are there than holly bushes?

If this door you wish to unlock

Upon which panels must you knock?

Door panels: 9 3 8 / 1 7 5 / 4 6 2

1 Jenny knocks on the 4 corner panels.
What is the total of these panel numbers?

2 Paul knocks on 3 panels.
Their numbers total **more than** 20.
Which panels could these be?

3 Ruth knocks on 4 panels.
They total **less than** 13.
Which panels could these be?

4 Ali knocks on the 5 panels which
have the **smallest** total.
What is this total?

5 To unlock the door the children must knock
on 4 panels which have a total of **exactly** 30.
Which panels must they knock on?

Go to Number Workbook page 1.

1 2 3 4 5 6 7 8 9

Copy and complete each magic shape
using the magician's numbers. Use
each number once only.

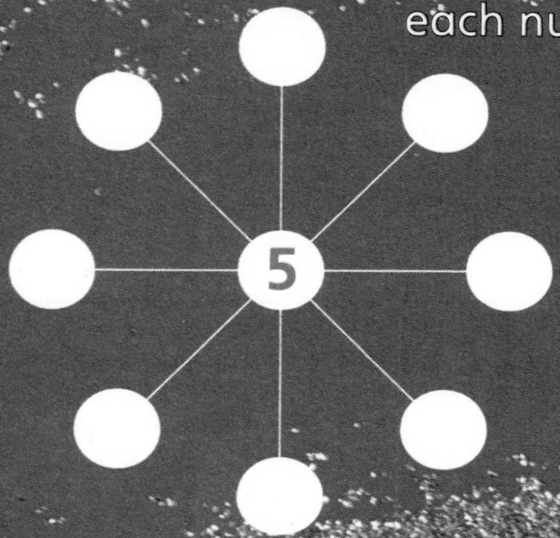

Problem solving

MAGIC CIRCLE

The 3 numbers in each
line must add to **15**.

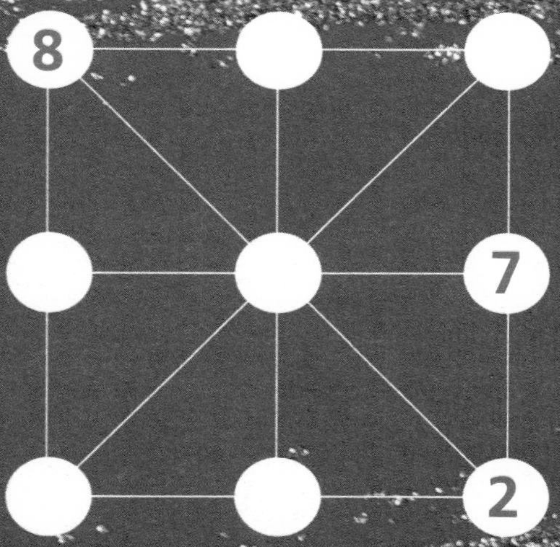

MAGIC SQUARE

8

7

2

The 3 numbers in each
line must add to **15**.

Extension

3

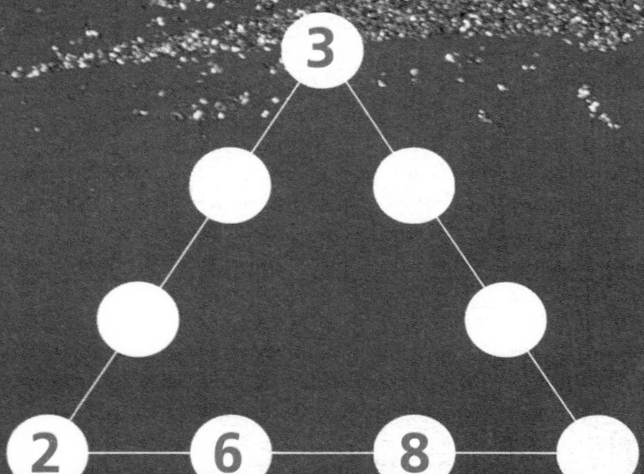

MAGIC TRIANGLE

The 4 numbers in each
line must add to **17**.

2 6 8

Mystery Manor attic

168

304

695

Use
hundreds
tens and
units if you
wish

The children find 4 keys.
On each label **one** answer matches a number
on an object in the attic.

Paul's key

1 (a) 8 2
        + 4 1
        _____

(b) 6 3
   + 6 2
   _____

(c) 8 4
   + 5 4
   _____

(d) 7 5
   + 6 2
   _____

(e) 9 3
   + 4 3
   _____

(f) 9 0
   + 9 0
   _____

(g) 7 0
   + 7 5
   _____

(h) 7 8
   + 9 0
   _____

(i) 2 2
   + 8 6
   _____

(j) 5 3
   + 5 6
   _____

2 Which object does Paul's key open?

Jenny's key

3 (a) 1 2 3
       + 3 2 1
       _____

(b) 4 2 5
   + 3 3 2
   _____

(c) 1 0 2
   + 2 3 2
   _____

(d) 4 0 1
   + 5 0 6
   _____

(e) 6 7 6
   +   2 3
   _____

(f) 5 0 5
   + 3 0 5
   _____

(g) 3 6 4
   + 4 2 9
   _____

(h) 9 1 3
   +   6 8
   _____

(i) 5 4 6
   + 4 2 4
   _____

(j) 5 2 8
   + 1 6 7
   _____

4 Which object does Jenny's key open?

803

900

777

Ali's key

5 (a) 174
+ 1 5 1

(b) 1 8 3
+ 1 6 4

(c) 2 7 2
+ 1 4 4

(d) 2 5 1
+   6 4

(e) 3 5 0
+ 2 7 2

(f) 2 4 3
+ 5 6 5

(g)   8 7
+ 5 3 1

(h) 2 7 3
+ 5 3 0

(i) 1 6 6
+ 5 5 1

(j) 6 9 0
+ 1 8 1

6 Which object does Ali's key open?

Ruth's key

7 (a) 1 4 8
+ 1 7 7

(b) 1 3 6
+ 1 7 7

(c) 1 6 6
+   6 8

(d) 3 7 9
+ 2 7 6

(e)   9 4
+ 4 2 7

(f) 4 6 9
+ 4 3 2

(g) 2 0 9
+ 5 9 3

(h) 5 9 8
+ 2 5 2

(i) 3 2 5
+ 5 7 5

(j)   6 5
+ 6 3 9

8 Which object does Ruth's key open?

---

9  Make up an addition for the key to the safe.

Extension

Turn the keys

Paul's key opens the wardrobe. It is full of toys.

1 There are 84 tin soldiers on the top shelf and 55 on the bottom.
How many soldiers are there altogether?

2 There are 87 marbles in the red bag and 115 in the green bag.
How many marbles are there altogether?

3

(a)
```
  9 9
+ 8 8
─────

─────
```

(b)
```
  4 9 1
+ 2 6 8
───────

───────
```

(c)
```
  5 5 5
+ 3 8 7
───────

───────
```

(d)
```
  4 9 6
+ 1 0 6
───────

───────
```

(e)
```
    6 6
+ 5 3 4
───────

───────
```

Jenny's key opens the suitcase. She finds a code.

A	D	E	H	I	L	M	P	T	S
9	8	7	6	5	4	3	2	1	0

4 Use the code to find the message.

(a)
```
  1 4 8
+   1 9
───────
  1 6 7
───────
  T H E
```

(b)
```
  1 8 5
+ 2 7 3
───────

───────
```

(c)
```
  3 9 8
+ 2 9 2
───────

───────
```

(d)
```
    5
+   4
─────

─────
```

(e)
```
  2 0 5
+ 1 8 7
───────

───────
```

5 (a) Find the total of 399 and 253.
   (b) What word is this?
   (c) Make up another word. Write a sum for it.

Extension

Ali's key opens the cupboard. It is full of comics and books.

1 One bundle has 117 comics.
The other has 99 comics.
How many comics are there altogether?

2 One shelf has 53 books.
The other has 49 books.
Find the total number of books.

3 (a) Find the sum of 349 and 506.
(b) What is 95 plus 312?
(c) Add 299 to 394.

Ruth's key opens the chest. She finds a strange maths book.

4 (a)
```
  193
+ 650
─────
```
(b)
```
  616
+ 227
─────
```
5 (a)
```
  824
+  97
─────
```
(b)
```
   60
+ 861
─────
```

(c)
```
  334
+ 509
─────
```
(d)
```
  270
+ 573
─────
```
(c)
```
  408
+ 435
─────
```
(d)
```
  639
+ 282
─────
```

(e)
```
   95
+ 826
─────
```
(f)
```
  513
+ 408
─────
```
(e)
```
  781
+  62
─────
```
(f)
```
  462
+ 459
─────
```

6 Check your answers. What do you notice?

**Ask your teacher
what to do next.**

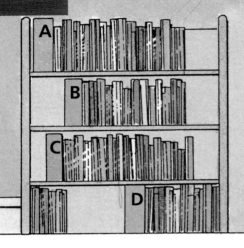

A B C D E F G H

328

Use hundreds tens and units if you wish

Some of the books on each shelf were damp and had to be thrown out.

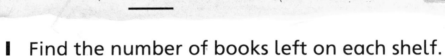

There were **3 5 1** on the **K** shelf.
           1 2 6 were thrown out.
           2 2 5 are left.

1 Find the number of books left on each shelf.

 A
```
  268
- 129
-----
  139
```

 B
```
  367
- 118
-----
```

C
```
  456
- 209
-----
```

D
```
  345
-  17
-----
```

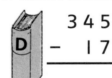 E
```
  690
- 338
-----
```

 F
```
  473
- 426
-----
```

 G
```
  582
- 364
-----
```

 H
```
  631
- 417
-----
```

I
```
  552
- 216
-----
```

 J
```
  281
-  32
-----
```

2 Find the difference between the number of books in the sacks.

139 books

253 books

Use hundreds tens and units if you wish

Some of the magazines were chewed by mice.

There were **4 1 6** magazines on the shelf.

1 5 1 were thrown out.

2 6 5 are left.

1 Find the number of magazines left.

(a)   3 8 6
    − 1 9 2
    _____

(b)   2 3 8
    − 1 8 0
    _____

(c)   3 2 9
    − 1 7 4
    _____

(d)   5 0 5
    − 2 1 4
    _____

(e)   4 1 7
    −   6 3
    _____

(f)   2 7 7
    − 1 9 1
    _____

(g)   2 4 9
    −   5 8
    _____

(h)   3 6 6
    − 1 7 6
    _____

(i)   4 0 9
    − 2 2 2
    _____

(j)   4 5 3
    − 3 9 0
    _____

2 Find the start of the secret passage to the cellar.

It is behind the picture whose number is

655 minus 372.

Which picture is it?

13 Into the cellar

Subtraction:
exchanging
a ten and a
hundred

MYSTERY MANOR

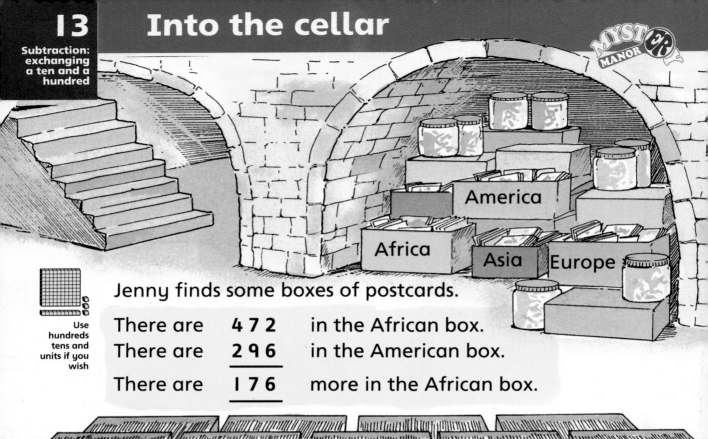

Use hundreds tens and units if you wish

Jenny finds some boxes of postcards.

There are **472** in the African box.
There are **296** in the American box.

There are **176** more in the African box.

1

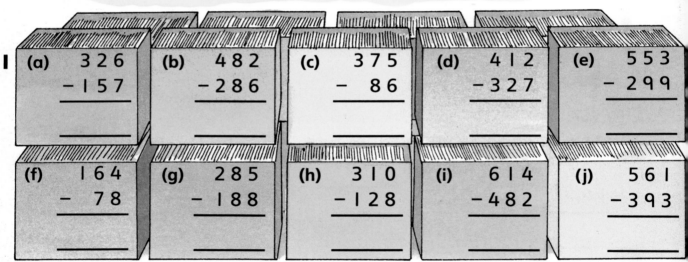

(a)
```
  326
- 157
-----
```

(b)
```
  482
- 286
-----
```

(c)
```
  375
-  86
-----
```

(d)
```
  412
- 327
-----
```

(e)
```
  553
- 299
-----
```

(f)
```
  164
-  78
-----
```

(g)
```
  285
- 188
-----
```

(h)
```
  310
- 128
-----
```

(i)
```
  614
- 482
-----
```

(j)
```
  561
- 393
-----
```

2 Ruth finds some jars of coins.

Germany 465
Australia 269
Brazil 168
India 321

(a) Find the difference between the number of coins in the German jar and the Australian jar.

(b) Which jar has more coins – the Brazilian jar or the Indian jar?

(c) How many more does it have?

Use hundreds tens and units if you wish

Paul finds an old cupboard with different items in the drawers.

There are **300** screws in one drawer.

There are **147** nails in another.

There are **153** more screws than nails.

1

(a)
```
   500
 - 242
 _____
```

(b)
```
   300
 - 183
 _____
```

(c)
```
   200
 -  96
 _____
```

(d)
```
   600
 - 325
 _____
```

(e)
```
   400
 - 214
 _____
```

(f)
```
   700
 - 357
 _____
```

(g)
```
   301
 - 283
 _____
```

(h)
```
   203
 -  58
 _____
```

(i)
```
   402
 - 323
 _____
```

(j)
```
   506
 -  89
 _____
```

419   269   380   480

2 Ali finds stacks of tiles on the floor.

(a) Find the difference between the number of blue tiles and the number of green tiles.

(b) How many more red tiles than yellow tiles are there?

(c) What is the difference between the number of tiles in the largest stack and the smallest stack?

In the stable

These four old cars have stopped here during a rally.

1 Find the difference between the numbers on
  (a) the blue and yellow cars
  (b) the blue and red cars
  (c) the green and yellow cars.

2 The number on the red car is 150 more than
the number on one of the other cars.
What is the colour of the other car?

3 At the end of the first day each car had
travelled these distances.

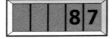 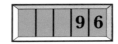  

How many miles further has
(a) the blue car travelled than the red car
(b) the yellow car travelled than the green car?

4 The log books show how far each car had
travelled at the end of the second day.

To complete the rally each car has to travel 400 miles.
How much further does each car have to travel?

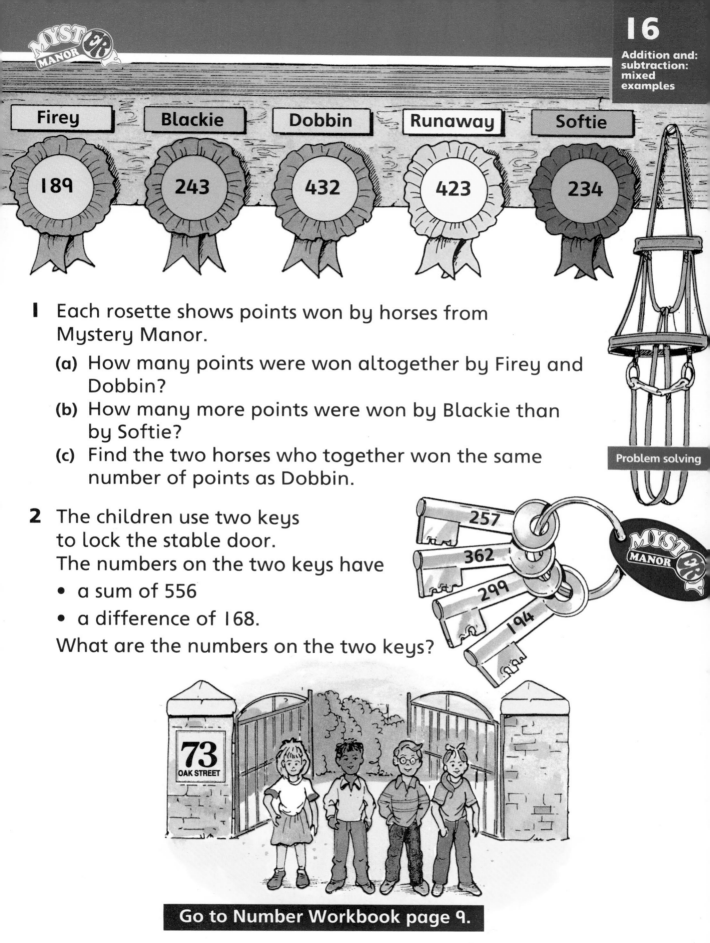

Firey	Blackie	Dobbin	Runaway	Softie
189	243	432	423	234

1 Each rosette shows points won by horses from Mystery Manor.

   (a) How many points were won altogether by Firey and Dobbin?

   (b) How many more points were won by Blackie than by Softie?

   (c) Find the two horses who together won the same number of points as Dobbin.

Problem solving

2 The children use two keys to lock the stable door.
The numbers on the two keys have

  • a sum of 556

  • a difference of 168.

What are the numbers on the two keys?

257
362
299
194

73
OAK STREET

Go to Number Workbook page 9.

Counting to one thousand

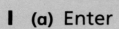 

nine hundred and ninety five . . .
nine hundred and ninety six . . .

I (a) Enter **996.**

Say and write the number
you see.

(b) Keep pressing **+** **I** **=**
Say and write the next **three** numbers.

(c) Add I more.
Copy and complete. 999 + I = _____

1000 is called **one thousand.**

960
950

2 (a) Enter **950.**

Press **+** **I** **0** **=**
Say and write the number you see.

(b) Keep pressing **+** **I** **0** **=**
Find the target by saying and writing
the next **four** numbers you see.

3 Enter **500.**

Keep pressing
+ **I** **0** **0** **=**
Say and write the
next **five** numbers.

five hundred . . .
six hundred . . .

Help
Childhelp
help
children

*Childhelp raises
money for children
in need*

CHILD Ch HELP

1 How many?

(a) 5 sets of 4 egg cups

(b) 2 boxes of 10 tumblers

(c) 4 bundles of 8 teaspoons

(d) 3 sets of 6 mugs

2 How many?

(a) 4 cards of 6 pens

(b) 4 packs of 9 'Childhelp Times'

(c) 5 sets of 7 pencil cases

(d) 2 sets of 8 rulers

3 How many?

(a) 3 cards of 5 badges

(b) 4 bundles of 7 posters

(c) 3 boxes of 8 cassettes

(d) 10 sheets of 8 stamps

4 Which shelf has most items?

Go to Number Workbook page 10.

38

38

Indira collected two envelopes
with 38 stamps in each.

She collected 76 stamps altogether.

Childhelp
We need
USED
STAMPS

$\begin{array}{r} 38 \\ \times 2 \\ \hline 76 \end{array}$

38

38

1 Multiply to find how many stamps are in
each pair of envelopes.

(a) 27
27

(b) 45
45

(c) 63
63

(d) 82
82

(e) 69
69

(f) 75
75

2 Multiply to find how many stamps are in
each set of three envelopes.

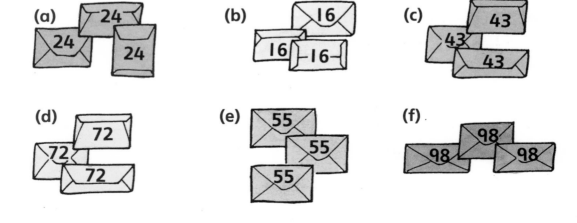

(a) 24
24
24

(b) 16
16 16

(c) 43
43
43

(d) 72
72
72

(e) 55
55
55

(f) 98
98 98

I Copy and complete each calculation from Indira's book.

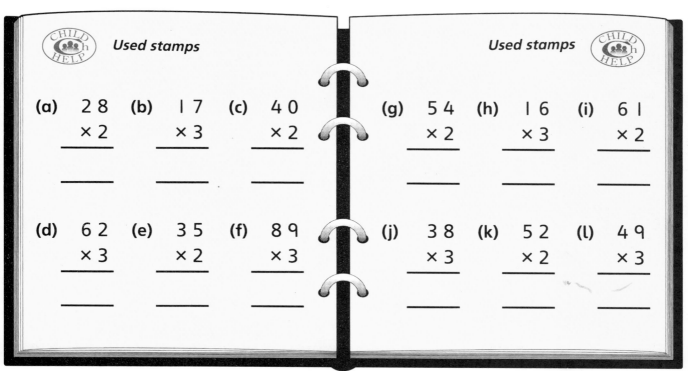

Used stamps

(a) 2 8
   × 2
_____

(b) 1 7
   × 3
_____

(c) 4 0
   × 2
_____

(g) 5 4
   × 2
_____

(h) 1 6
   × 3
_____

(i) 6 1
   × 2
_____

(d) 6 2
   × 3
_____

(e) 3 5
   × 2
_____

(f) 8 9
   × 3
_____

(j) 3 8
   × 3
_____

(k) 5 2
   × 2
_____

(l) 4 9
   × 3
_____

Used stamps

2 Multiply to find the total number of stamps in
   (a) the blue envelopes
   (b) the green envelopes
   (c) the yellow envelopes.

3 How many stamps are there altogether in the blue, green and yellow envelopes?

Problem solving

4 The total number of stamps in all ten envelopes is 296. Each red envelope has the same number of stamps. How many stamps are in each red envelope?

1 Leela ordered 4 boxes of each of these.
How many of each item did she order?

(a)
$$\begin{array}{r} 12 \\ \times\ 4 \\ \hline \end{array}$$

(b)
$$\begin{array}{r} 18 \\ \times\ 4 \\ \hline \end{array}$$

(c)
$$\begin{array}{r} 24 \\ \times\ 4 \\ \hline \end{array}$$

(d)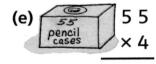
$$\begin{array}{r} 32 \\ \times\ 4 \\ \hline \end{array}$$

(e)
$$\begin{array}{r} 55 \\ \times\ 4 \\ \hline \end{array}$$

(f)
$$\begin{array}{r} 96 \\ \times\ 4 \\ \hline \end{array}$$

2 (a) 4×81 (b) 4×79 (c) 4×63 (d) 4×48
   (e) 4×60 (f) 4×77 (g) 4×95 (h) 4×19

3 She ordered 5 boxes of each of these.
How many of each item did she order?

(a)
$$\begin{array}{r} 11 \\ \times\ 5 \\ \hline \end{array}$$

(b)
$$\begin{array}{r} 18 \\ \times\ 5 \\ \hline \end{array}$$

(c)
$$\begin{array}{r} 36 \\ \times\ 5 \\ \hline \end{array}$$

(d)
$$\begin{array}{r} 22 \\ \times\ 5 \\ \hline \end{array}$$

(e)
$$\begin{array}{r} 27 \\ \times\ 5 \\ \hline \end{array}$$

(f)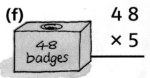
$$\begin{array}{r} 48 \\ \times\ 5 \\ \hline \end{array}$$

4 (a) 5×63 (b) 5×40 (c) 5×99 (d) 5×56
   (e) 5×84 (f) 5×92 (g) 5×75 (h) 5×88

Special offer

Number of tokens needed

badge
35 tokens

book
99 tokens

pen
19 tokens

cassette
77 tokens

mug
87 tokens

pencil
12 tokens

calculator
94 tokens

game
50 tokens

comb
23 tokens

torch
81 tokens

1 How many tokens are needed for

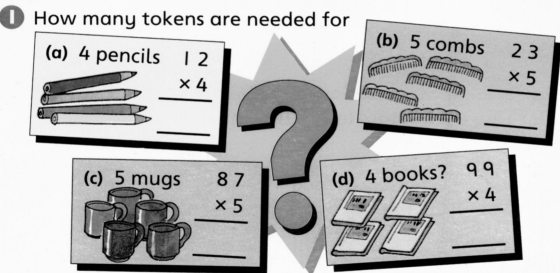

(a) 4 pencils

$$\begin{array}{r} 1\ 2 \\ \times\ 4 \\ \hline \end{array}$$

(b) 5 combs

$$\begin{array}{r} 2\ 3 \\ \times\ 5 \\ \hline \end{array}$$

(c) 5 mugs

$$\begin{array}{r} 8\ 7 \\ \times\ 5 \\ \hline \end{array}$$

(d) 4 books?

$$\begin{array}{r} 9\ 9 \\ \times\ 4 \\ \hline \end{array}$$

2 How many tokens are needed for

- **(a)** 5 calculators
- **(b)** 4 badges
- **(c)** 5 cassettes
- **(d)** 4 pens
- **(e)** 5 torches
- **(f)** 4 games?

3 How many tokens are needed for

- **(a)** 5 pencils and 4 cassettes
- **(b)** 4 mugs and 5 books?

4 Make up multiplication questions using the tokens chart.
Check with a calculator.

Ask your teacher what to do next.

CHILD HELP

One way to add on 9 is to **add 10 and take I off.**

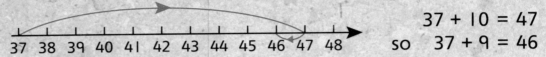

37 38 39 40 41 42 43 44 45 46 47 48

$$37 + 10 = 47$$
so $37 + 9 = 46$

I Add 9 to each of these numbers.

(a) 32   (b) 56   (c) 49   (d) 94   (e) 125   (f) 108

10 20 50

Alan's total was 50.
He scored 20, 20 and 10 with
his three darts.

2 What do you think these
players scored with
each of their three darts?

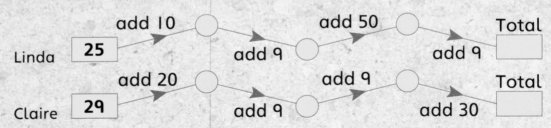

Claire
Total 90

Linda
Total 110

Scott
Total 120

3 **(a)** Find each total. Record like this: Linda 25, 35, ...

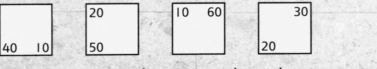

Linda **25** — add 10 → ○ — add 9 → ○ — add 50 → ○ — add 9 → Total ▢

Claire **29** — add 20 → ○ — add 9 → ○ — add 9 → ○ — add 30 → Total ▢

(b) Whose total is greater? How much greater?

4 You need four squares. Write these numbers on them.

40 10   20   10 60   30
              50   20

Put them together to make a larger square.
It must have four numbers in the middle
which **add to 100.**
Draw your answer.

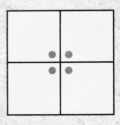

In the box there are
6 rows of 4 cakes.

$$6 \times 4 = 24$$

There are 24 cakes
altogether.

1 How many cakes are there on each of these trays?

(a) 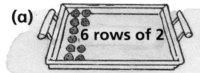 6 rows of 2

(b)  6 rows of 6

(c)  6 rows of 9

2

Jan brought 6 plates
each with 7 scones.
Tim brought 5 plates
each with 8 scones.

(a) Who brought more scones?
(b) How many more?

3 Copy and complete this sequence from the 6 times table.

0, 6, 12, _____ , _____ , 30, _____ , _____ , 48, _____ , 60

4 (a) $(6 \times 1) + 3$ (b) $(6 \times 3) - 8$ (c) $(6 \times 10) + 5$

5 (a) $6 \times \boxed{} = 24$ (b) $6 \times \boxed{} = 60$ (c) $6 \times \boxed{} = 42$

6 In the hall, there are
6 tables each with 4 chairs
6 tables each with 5 chairs
6 tables each with 8 chairs

Are there enough chairs for
100 people?
Explain your answer.

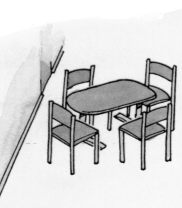 

1 Ken made up 6 plates of each kind of sandwich.
How many of each kind did he make?

(a)

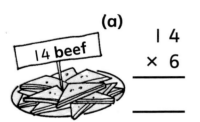

14 beef

    1 4
  × 6
  _____

(b)

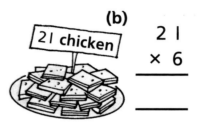

21 chicken

    2 1
  × 6
  _____

(c)

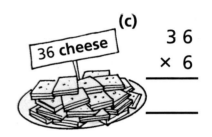

36 cheese

    3 6
  × 6
  _____

2 **(a)** 6 × 30   **(b)** 6 × 27   **(c)** 6 × 96   **(d)** 6 × 38
    (e) 44 × 6   **(f)** 6 × 51   **(g)** 6 × 60   **(h)** 39 × 6

3

To Childhelp
Pears – 6 boxes of 18
Apples – 6 boxes of 32
Oranges – 6 boxes of 50
Peaches – 6 boxes of 24

Some fruit was given
to Childhelp.

How many of each fruit was given:
(a) pears   **(b)** apples
(c) oranges   **(d)** peaches

4

14p each

pears

16p each

apples

15p each

oranges

Ann had £1. She bought 6 of one type of fruit
and was given 10p change. Which fruit did she buy?

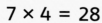

There are 7 cars with
4 people in each.

$7 \times 4 = 28$

There are 28 people altogether.

1 How many are there altogether?

(a) 7 lots of 2 cans

(b) 7 rows of 5 stickers

(c) 7 rows of 6 pins

(d) 7 lots of 10 pennies

2 **(a)** 7×3 **(b)** 7×8 **(c)** 7×7 **(d)** 7×9

3 **(a)** $(7 \times 6) + 1$ **(b)** $(7 \times 10) - 5$ **(c)** $7 \times \boxed{} = 42$

4

How many of each
of these things will a
collector bring in 7 days?

Instructions for collectors

Each day, bring with you

- 1 biscuit
- 7 teabags
- 4 sandwiches
- 9 pence

27

Multiplication:
the 7 times
table

★ Competitions ★ Comp

1 Copy and complete this sequence from the 7 times table.

▷ ▶ 0, 7, 14, ___ , 28 , ___ , ___ , ___ , 56 , ___ , 70 ◀ ◀

2 A number from the 7 times table wins a badge.

56 wins because 7 × 8 = 56.

Which of these numbers will win and why?

Pick a number

56   43   20   24   29   32   10

15   49   45   21

60   65   54   30   37   35

3 This is a 7 times machine.

When 6 is put **IN**, 42 comes **OUT**

What **OUT** number should
each person guess?

9   0   8   4

Lynne   Ray   Chris   Joy

IN ▷   7 times ▷   OUT ▷

CHILD h HELP

4 A card wins if
one row or
column has all
of its numbers in
the 7 times table.

★ Three in a line ★

4	36	70
35	54	49
12	19	14

42	52	7
63	20	48
32	27	16

30	45	3
28	56	21
18	24	12

(a) Which cards will win? Why?
(b) Make up your own card which will win.

Problem solving

itions ★ Competitions ★

1 Childhelp ordered 7 boxes of each of these.
How many items are ordered?

96 puzzle books

(a)
```
  9 6
× 7
─────
```

72 pens

(b)
```
  7 2
× 7
─────
```

2 (a) Each box of puzzle books costs £24.
How much does 7 boxes cost?

(b) Each box of pens costs £13.
How much does 7 boxes cost?

(c) What is the total cost?

WIN
a prize for
answers between
300 and 600

77777777777777777777777777777777

3 Childhelp Times offers a prize.
Multiply to see if you can win a prize.

(a)
```
  2 9
× 7
─────
```

(b)
```
  8 6
× 7
─────
```

(c)
```
  6 0
× 7
─────
```

(d)
```
  5 4
× 7
─────
```

(e)
```
  9 1
× 7
─────
```

(f)
```
  1 6
× 7
─────
```

(g)
```
  4 3
× 7
─────
```

(h)
```
  5 0
× 7
─────
```

(i)
```
  9 8
× 7
─────
```

(j)
```
  8 5
× 7
─────
```

77777777777777777777777777777777

4 (a) 7 × 52 (b) 7 × 65 (c) 7 × 79 (d) 7 × 33
(e) 83 × 7 (f) 75 × 7 (g) 7 × 61 (h) 97 × 7

5 Childhelp collects money for 17 weeks.
How many days is this?

Judy's Octopus team

There are 8 members in Judy's team.
They each have 3 T-shirts.

$$8 \times 3 = 24$$

They have 24 T-shirts altogether.

1 Each member of the team of 8
has these items.

(a) 4 pairs of shorts   (b) 5 arm bands
(c) 2 track suits   (d) 9 medals
(e) 7 pairs of socks   (f) 6 running vests

Find the total number of each item.

2 (a) $(8 \times 3) + 1$ (b) $(8 \times 8) - 1$ (c) $8 \times \boxed{} = 56$

(d) $8 \times \boxed{} = 80$ (e) $(8 \times 6) - 6$ (f) $8 \times \boxed{} = 72$

Target
£300

Each member of the team ran 8 laps.

3 Judy collected £10 for each of her laps.
How much did she collect altogether?

4 The other team members collected these
amounts for each lap:
(a) Leela £8   (b) Irfan £4   (c) Colin £7   (d) Josh £1
(e) Les £5   (f) Suzy £9   (g) Gina £3

How much did each member collect altogether?

5 The team set a target of £300.
Did they make their target?

Run 8 laps for Childhelp.

1 Judy's Octopus team bounced for Childhelp. All 8 team members bounced 31 times each.

Find the total number of bounces for Judy's team.

```
  3 1
× 8
____
```

2 Find the total number of bounces for each of these teams.

(a) Dolphins
```
  2 8
× 8
____

____
```

(b) Whales
```
  6 3
× 8
____

____
```

(c) Otters
```
  4 9
× 8
____

____
```

(d) Seals
```
  7 6
× 8
____

____
```

3 (a)
```
  3 5
× 8
____

____
```
(b)
```
  9 1
× 8
____

____
```
(c)
```
  8 8
× 8
____

____
```
(d)
```
  7 0
× 8
____

____
```
(e)
```
  8 2
× 8
____

____
```

4 (a) 8×37 (b) 8×66 (c) 8×75 (d) 8×99

sticker **86p**

pennant **55p**

headband **67p**

medal **59p**

5 Jason has £5.
He wants to buy 8 of the same item for the Puffin team.
Which item could it be?

Sponsored Swim

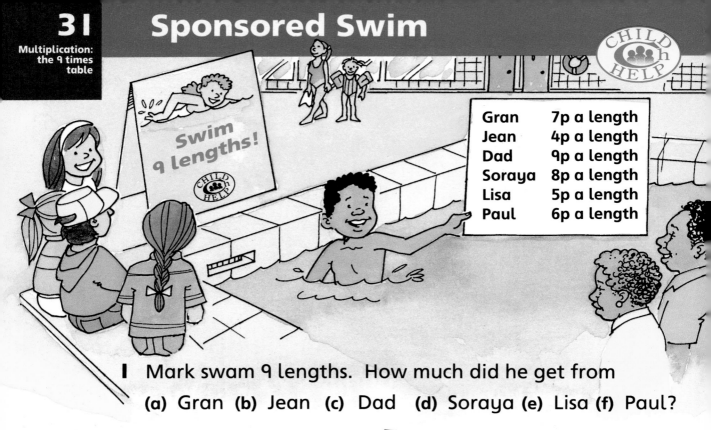

Gran	7p a length
Jean	4p a length
Dad	9p a length
Soraya	8p a length
Lisa	5p a length
Paul	6p a length

1 Mark swam 9 lengths. How much did he get from

(a) Gran (b) Jean (c) Dad (d) Soraya (e) Lisa (f) Paul?

Problem solving

2 Rosie swam 9 lengths.

How much did
each sponsor give
per length?

		total collected
(a)	Grandpa	45p
(b)	Emma	27p
(c)	Aunt Kate	63p

Problem solving

3 These amounts were raised by swimming 9 lengths:
63p, 45p, 81p, 90p, 32p, and 54p.

Mark knew that one amount was wrong.
Which is it and how did he know?

4 (a) $(9 \times 1) - 1 =$ _____

$(9 \times 2) - 2 =$ _____

$(9 \times 3) - 3 =$ _____

$(9 \times 4) - 4 =$ _____

$(9 \times 5) - 5 =$ _____

$(9 \times 6) - 6 =$ _____

(b) $(9 \times 1) + 1 =$ _____

$(9 \times 2) + 2 =$ _____

$(9 \times 3) + 3 =$ _____

$(9 \times 4) + 4 =$ _____

$(9 \times 5) + 5 =$ _____

$(9 \times 6) + 6 =$ _____

(c) What do you notice about your answers?

1 8
× 9

1 Mark took 18 strokes to swim a length.
How many strokes did he take for 9 lengths?

2 How many strokes did each swimmer take for 9 lengths?

(a) Rosie **(b)** Duncan **(c)** Kim

24 strokes
each length

17 strokes
each length

26 strokes
each length

3 Find the total number of lengths for each school.

Name of school	Number who swam	Lengths	Total lengths
Dean Park	28	9	
Fern Road	35	9	
Gordon	42	9	
Lowlands	29	9	
Taylors	37	9	

4 **(a)** 9 × 51 **(b)** 9 × 68 **(c)** 9 × 34 **(d)** 9 × 99

5 **(a)** Write the missing numbers for the blue and yellow ladders.

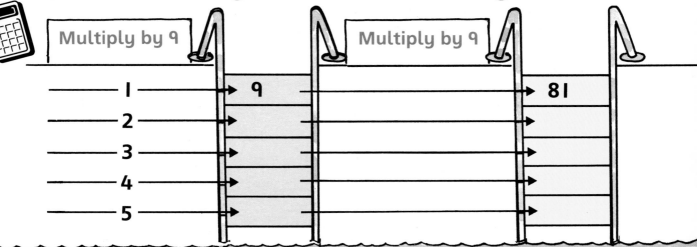

Multiply by 9 Multiply by 9

1 → 9 → 81
2 → →
3 → →
4 → →
5 → →

(b) What patterns do you see on the yellow ladder?

Sarah's problem page

1 Write $7 \times 2 = 14$ using words only.

2

(a) Which of these could be solved by doing a multiplication?

A

John has 6p more than Sarah. How much does he have?

Sarah has 27p.

B

Dave has twice as much as Sarah. How much has he?

C

Ann has 12p less than Sarah. How much has she?

D

Sheena has three times as much as Sarah. How much has she?

E

Steve has double the amount of money Sarah has. How much has he?

(b) Which two children have the same amount of money?

3 Copy and complete these words used in multiplication:

(a) m u _ _ _ _ _ l y   **(b)** t _ _ _ s   **(c)** d _ _ _ _ _ e

4 Here is a word problem for 8×6. 👉

Write a word problem for

(a) 6×10 **(b)** 9×5.

There are 6 cakes in a box. How many are in 8 boxes?

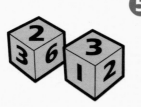

5 Make up rules for a multiplication game where the players use two dice.

Play your game.

CHILD HELP

Alan, Claire, Linda and Scott are playing an electronic game.
They press buttons to add and subtract numbers.

1 Which number would light up for each player
if these buttons were pressed?

(a) **add** **7**

(b) **subtract** **8**

(c) **add** **11**

2 (a) **Who** should be
happy to press **subtract** **5** Why?

(b) **Who** would be
unhappy? Why?

3 Which buttons should Scott press to be happy? Why?

4 Which buttons should Alan press to make 72 light up?

5 Which buttons should Claire press to make 83 light up?

Extension

You need a set of cards 1 2 3 4 5 6 7 8 9 .

Problem solving

1 (a) Use these four cards 4 5 6 7 .

Arrange the cards like this
to make a subtraction. ⟶

$$\begin{array}{r} 6\ 5 \\ -\ 4\ 7 \\ \hline \end{array}$$

Write
$$\begin{array}{r} 6\ 5 \\ -\ 4\ 7 \\ \hline \end{array}$$

(b) Using the cards, find and write
all the possible subtractions.

(c) Use a calculator to find
• the largest answer
• the smallest answer.

2 (a) Use these four cards 2 3 4 5 .

Find and write the subtractions
which have
• the largest answer
• the smallest answer.

$$\begin{array}{r} \square\ \square \\ -\ \square\ \square \\ \hline \end{array}$$

(b) What do you notice about
these answers and the
answers in question 1(c)?

> 4, 5, 6, 7 and 2, 3, 4, 5 are called **consecutive numbers**.

Extension

3 Choose a different set of four **consecutive numbers**.
Write the subtractions which give
• the largest answer
• the smallest answer.

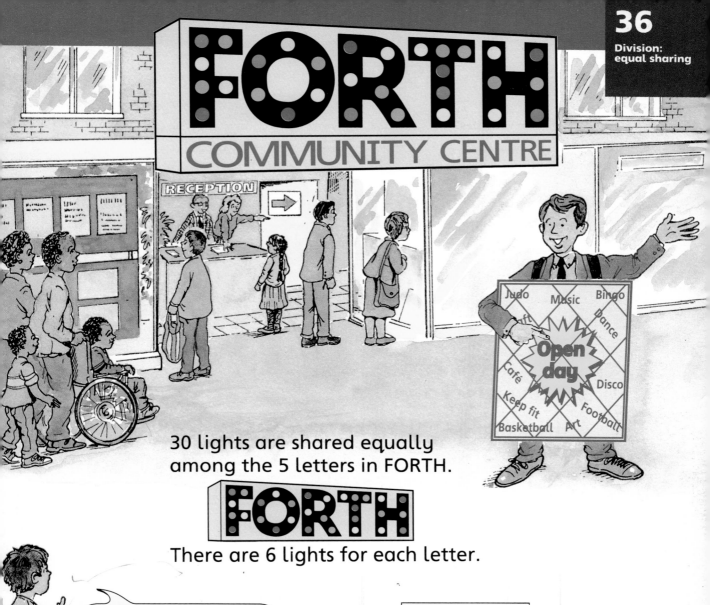

30 lights are shared equally among the 5 letters in FORTH.

There are 6 lights for each letter.

30 divided by 5 is 6   We write $30 \div 5 = 6$

Use counters or cubes. Write each division.

I **(a)** Divide 18 lights equally
among these 3 letters.

(b) Divide 15 lights equally
among these 5 letters.

(c) Divide 24 lights equally
among these 4 letters.

(d) Divide 20 lights equally
among these 10 letters.

Use counters or cubes.

I How many lights will there be on each letter of
these signs?

(a) WAY IN
25 lights

(b) GYM
24 lights

(c) HALL
28 lights

(d) TELEPHONES
30 lights

2 Copy and complete.

(a) $16 \div 2$ (b) $27 \div 3$ (c) $16 \div 4$ (d) $35 \div 5$

(e) $20 \div 5$ (f) $36 \div 4$ (g) $40 \div 10$ (h) $21 \div 3$

(i) $30 \div 3$ (j) $32 \div 4$ (k) $20 \div 2$ (l) $30 \div 5$

Problem solving

3 32 lights can be divided equally among
the letters of one of these signs. Which sign?

 WAY OUT    DISCO    CAFE

4 Draw a sign in which 12 lights can be shared
equally among the letters.

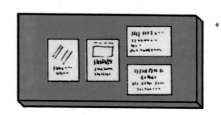

13 notices are to be shared equally among 3 boards.

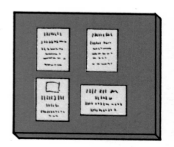

There are 4 notices on each board and 1 notice left over.

13 divided by 3 is 4 remainder 1

We write $13 \div 3 = 4 \text{ r } 1$

Use counters or cubes.

1 28 posters are divided equally into 3 piles.
  (a) How many are in each pile?
  (b) How many are left over?

2 Write each division.
  (a) Divide 23 pins equally among 5 notices.
  (b) Divide 19 pens equally among 2 helpers.
  (c) Divide 25 photographs equally among 4 displays.

3 Copy and complete.
  (a) $17 \div 3$ (b) $11 \div 5$ (c) $19 \div 10$ (d) $25 \div 3$
  (e) $23 \div 4$ (f) $13 \div 2$ (g) $18 \div 5$ (h) $21 \div 4$

4 Can two boys share 15 posters equally between them?

Explain your answer.

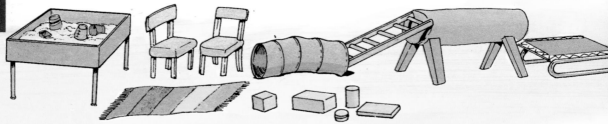

20 cars are in groups of 5.

There are 4 groups.

20 divided by 5 is 4   We write $20 \div 5 = 4$

Use counters or cubes. Write each division.

1 (a) There are 30 bricks.
How many towers of 5 can be made?

(b) 36 books are in a box.
Divide them into sets of 4.

(c) There are 18 children in the hall.
Divide them into teams of two.

2 How many groups can take part in these activities?

(a)
Singing
35 children
groups of 5

(b)
Water play
27 children
groups of 3

(c)
Sand play
16 children
groups of 4

(d)
Painting
40 people
groups of 10

3 Copy and complete.

(a) $20 \div 2$ (b) $40 \div 5$ (c) $32 \div 4$ (d) $30 \div 3$

(e) $60 \div 10$ (f) $15 \div 5$ (g) $14 \div 2$ (h) $20 \div 4$

The children make trains. 14 boxes are put in groups of 4.

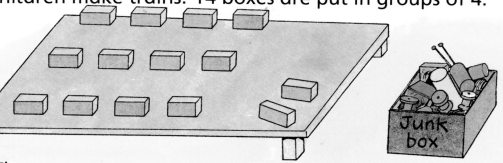

There are 3 groups and 2 left over.

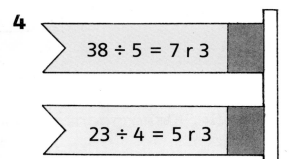

**14 divided by 4 is
3 remainder 2**

We write $14 \div 4 = 3 \text{ r } 2$

Use cubes or counters.

1 Divide 17 sticky shapes into groups of 3.
 (a) How many groups are there?
 (b) How many are left over?

2 Write each division.
 (a) Divide 19 tubes into groups of 2.
 (b) Divide 23 children into groups of 5 to make trains.
 (c) Divide 18 wheels into groups of 4.

3 Copy and complete.
 (a) $19 \div 4$ (b) $13 \div 2$ (c) $17 \div 5$ (d) $22 \div 3$
 (e) $24 \div 10$ (f) $29 \div 4$ (g) $25 \div 3$ (h) $33 \div 10$

4

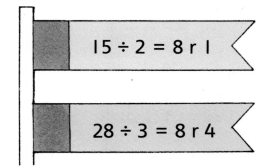

$38 \div 5 = 7 \text{ r } 3$

$15 \div 2 = 8 \text{ r } 1$

$23 \div 4 = 5 \text{ r } 3$

$28 \div 3 = 8 \text{ r } 4$

(a) Which of the signals show the wrong answers?
(b) Write the correct answers.

10 chairs are to be shared equally between 2 rooms.

10 divided by 2 . . .
2 times **what** is 10?

2 times **5** is 10

$10 \div 2 = 5$

There are 5 chairs for each room.

1 Divide these equally between 2 rooms.
- **(a)** 8 pictures
- **(b)** 2 tables
- **(c)** 14 plants
- **(d)** 16 pin boards
- **(e)** 20 stools
- **(f)** 6 shelves

2 Copy and complete.
- **(a)** $4 \div 2$
- **(b)** $18 \div 2$
- **(c)** $12 \div 2$
- **(d)** $10 \div 2$

12 mugs are to be divided equally among 3 trays.

3 times what is 12?

3 times 4 is 12

$12 \div 3 = 4$

There are 4 mugs on each tray.

3 Divide these equally among 3 trays.
- **(a)** 24 plates
- **(b)** 3 teapots
- **(c)** 6 sugar bowls
- **(d)** 18 cups
- **(e)** 15 saucers
- **(f)** 30 teaspoons

4 Copy and complete.
- **(a)** $9 \div 3$
- **(b)** $21 \div 3$
- **(c)** $12 \div 3$
- **(d)** $27 \div 3$
- **(e)** $14 \div 2$
- **(f)** $18 \div 3$
- **(g)** $24 \div 3$
- **(h)** $20 \div 2$

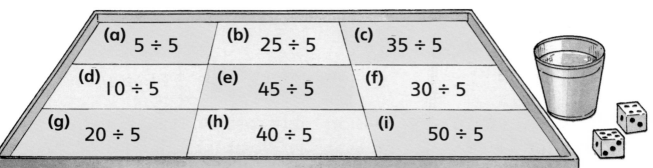

12 disc boxes are to be shared
equally among 4 trolleys.

4 times what is 12?   12 ÷ 4 = 3

There are 3 disc boxes for each trolley.

1 Divide these equally among the 4 trolleys.
  (a) 8 disc drives   **(b)** 20 programs   **(c)** 40 discs
  (d) 4 printers   **(e)** 24 books   **(f)** 16 sheets of paper

2 Copy and complete.

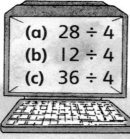
  (a) 28 ÷ 4
  (b) 12 ÷ 4
  (c) 36 ÷ 4

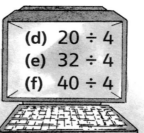
  (d) 20 ÷ 4
  (e) 32 ÷ 4
  (f) 40 ÷ 4

3

Divide these equally among 5 games tables.
  (a) 10 packs of cards   **(b)** 25 chess sets   **(c)** 15 puzzles
  (d) 20 shakers   **(e)** 30 counters   **(f)** 40 dice

4 Calculate.

(a) 5 ÷ 5	**(b)** 25 ÷ 5	**(c)** 35 ÷ 5
(d) 10 ÷ 5	**(e)** 45 ÷ 5	**(f)** 30 ÷ 5
(g) 20 ÷ 5	**(h)** 40 ÷ 5	**(i)** 50 ÷ 5

1 Divide these equally among 10 tables.

(a) 40 stools   (b) 50 glasses
(c) 10 jugs   (d) 60 plates
(e) 20 menus   (f) 100 napkins

2 Copy and complete.

(a) $40 \div 10$ (b) $60 \div 10$ (c) $90 \div 10$
(d) $30 \div 10$ (e) $70 \div 10$ (f) $80 \div 10$

There are 15 biscuits, 3 on each plate.

We can write $15 \div 3$ as $3\overline{)15}$

We write the answer here ⟶ $\dfrac{5}{3\overline{)15}}$

There are 5 plates of biscuits.

3 Do these in the same way.

(a) $32 \div 4$ (b) $40 \div 5$ (c) $27 \div 3$ (d) $50 \div 10$

4 Copy and complete.

(a) $3\overline{)18}$ (b) $4\overline{)20}$ (c) $2\overline{)14}$ (d) $5\overline{)30}$

(e) $10\overline{)20}$ (f) $3\overline{)24}$ (g) $4\overline{)24}$ (h) $2\overline{)16}$

(i) $5\overline{)25}$ (j) $4\overline{)36}$ (k) $3\overline{)21}$ (l) $5\overline{)35}$

5 (a) Copy and complete this division table up to $20 \div 2 = 10$.

$2 \div 2 = 1$
$4 \div 2 = 2$
$6 \div 2 = 3$

(b) Make up the division table for 3, 4, 5 **or** 10.

Go to Number Workbook page 18.

1 Share these items equally.
   (a) 18 bands among 3 girls
   (b) 14 hoops between 2 boys
   (c) 25 flags among 5 pupils
   (d) 30 balls among 10 girls

2 Group the children into teams.
   How many teams are there?
   (a) 18 children in teams of 2
   (b) 28 children in teams of 4

3 How many 5-a-side teams
   can be made from 35 players?

4 Teams of 3 were made from 29 parents.
   How many parents were **not** in a team?

5 Which of these can be done by division?

A Share 45 balls equally among 5 teams.

B Find the total number of skittles in 3 boxes of 18.

C There are 32 boys and 8 girls. How many more boys are there?

D There are 4 five-a-side teams. How many shirts do they need?

E Fifty flags are put in bundles of 10. How many bundles are there?

Problem solving

6 Which numbers between 30 and 40 have
   remainder 3 when divided by 4?

Use tens and units

Use tens and units.

1 Share these things equally among the 3 barbecues.
How many are there for each barbecue?

(a) 39 burgers

(b) 63 sausages

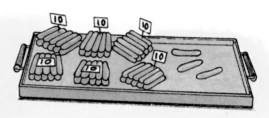

(c) 68 kebabs **(d)** 93 fish fingers **(e)** 37 onions

(f) 97 tomatoes **(g)** 65 mushrooms **(h)** 60 potatoes

2 Copy and complete.

(a) 64 ÷ 3 **(b)** 46 ÷ 2 **(c)** 60 ÷ 2 **(d)** 28 ÷ 2

(e) 25 ÷ 2 **(f)** 41 ÷ 2 **(g)** $2\overline{)47}$ **(h)** $3\overline{)35}$

(i) $4\overline{)45}$ **(j)** $4\overline{)48}$ **(k)** $5\overline{)56}$ **(l)** $5\overline{)59}$

3 Divide the rolls equally among the 3 barbecues.

33 ROLLS 33 ROLLS

4 The cook shared 34 loaves and 62 vegeburgers equally among the 3 barbecues.
What was left over for her to eat?

Use tens and units.

1 Share these things equally between the 2 tables.

(a) 32 bags of nuts

(b) 38 yoghurts

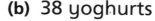

(c) 35 plates   (d) 54 apples   (e) 81 biscuits

2 Divide equally.   (a) 42 sandwiches among 3 plates
(b) 48 cans of juice among 3 boxes
(c) 40 oranges among 3 bowls
(d) 73 mugs among 3 trays
(e) 51 ice lollies among 3 cool boxes

3 Copy and complete.

(a) 33 ÷ 2   (b) 45 ÷ 2   (c) 98 ÷ 3   (d) 50 ÷ 4

(e) 64 ÷ 5   (f) 2)55   (g) 2)27   (h) 3)46

(i) 4)59   (j) 4)86   (k) 5)77   (l) 5)68

4 Choose **three** of these boxes to take to the barbecue.

cartons of milk   oranges   lollies   cans of coke

Share the food equally among 3 groups.
Write a list of what each group gets.

The quizmaster shared 53 quiz sheets equally between the 2 tables.

There are 26 quiz sheets on each table and 1 quiz sheet left over.

1 Divide equally.

(a) 47 pencils between 2 tables

(b) 49 score sheets among 3 helpers

(c) 39 pens between 2 tables

(d) 57 cards among 3 helpers

2 (a) $2\overline{)86}$ (b) $3\overline{)87}$ (c) $2\overline{)61}$ (d) $3\overline{)64}$

(e) $2\overline{)49}$ (f) $3\overline{)98}$ (g) $2\overline{)99}$ (h) $3\overline{)50}$

(i) $80 \div 3$ (j) $55 \div 2$ (k) $77 \div 3$ (l) $70 \div 2$

3 How many sets of 3 score sheets can be made from a bundle of 72?

4 Divide 58 cups of tea between 2 tables. How many cups of tea are on each table?

5 An envelope holds 3 answer books. How many envelopes are needed for 59 answer books?

1 Divide 60 singers equally into 4 rows.
How many singers are in each row?

2 Divide these equally among 4 music groups.

(a) 48 flutes
(b) 64 trumpets
(c) 98 music stands
(d) 59 violins

3 Divide 60 children equally into 5 music groups.
How many are in each group?

4 Divide these equally among the 5 groups.

(a) 70 triangles   (b) 85 drums   (c) 67 beaters
(d) 75 shakers   (e) 84 chime bars   (f) 55 bells

5 Copy and complete.

(a) $4\overline{)52}$ (b) $5\overline{)90}$ (c) $5\overline{)63}$ (d) $4\overline{)61}$

(e) $5\overline{)95}$ (f) $4\overline{)73}$ (g) $4\overline{)87}$ (h) $5\overline{)78}$

(i) $68 \div 4$ (j) $76 \div 5$ (k) $71 \div 4$ (l) $80 \div 5$

6 The number on this programme is a lucky number.
It divides by 4 **and has no remainder**.

(a) Use cards with the digits 6 7 8 9 .
Make at least six different programme numbers
each less than 100.

(b) Which of these numbers are lucky numbers?

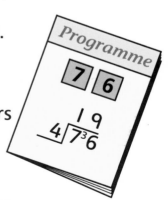

Programme

7 6

$4\overline{)7^36}$ = 1 9

7 Arrange 50 music stands in groups of 4. How many
groups are there?

Basketball match

ORTH
COMMUNITY
CENTRE

I Match the balls to the nets like this:

A	B	C
$77 \div 7 = 11$		

odd numbers
less than 14

even numbers
less than 15

numbers
greater than 15

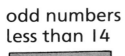

A

B

C

 72 ÷ 3

36 ÷ 3

90 ÷ 5

 77 ÷ 7

 96 ÷ 8

 96 ÷ 6

 98 ÷ 2

  90 ÷ 10

 85 ÷ 5

  48 ÷ 3

  92 ÷ 4

  99 ÷ 9

  54 ÷ 2

 98 ÷ 7

  84 ÷ 6

  65 ÷ 5

  84 ÷ 7

2 Draw other balls to match each net.

Ask your teacher what to do next.

Problem solving

You need counters. Work with a partner.

1 (a) Use 12 counters.
   Make 2 equal rows
   of 6 counters like this.
   Write 12 = 2 × 6

   (b) Find other ways of making equal rows using 12 counters.

   For each way write 12 = _____ × _____

2 Use 20 counters.
   Make equal rows using 20 counters.

   Copy and complete this table
   to show all the numbers that
   ☐ and △ could be.

20 =	☐	×	△
20 =	4	×	5
20 =			

You may use a calculator.

3 Copy and complete each table
   to show all the numbers that ☐ and △ could be.

(a)

24 =	☐	×	△
24 =	3	×	
24 =	2	×	

(b)

36 =	☐	×	△
36 =			
36 =			

4 List all the pairs of numbers which multiply to give
   (a) 18   **(b)** 30   **(c)** 48

Bouncer and Spot find their way home.
All the numbers on Bouncer's path **divide exactly** by 2.
All the numbers on Spot's path **divide exactly** by 5.

1 (a) Trace out Bouncer's path home to find where he lives.

(b) Write down all the **circled** numbers on his path.

2 Do all this again for Spot.

3 Going home the two dogs sometimes pass
through the **same** circled numbers.

(a) Write down these numbers.

(b) What do you notice about them?

Green Farm   White Gables   Rose Cottage   Trent Lodge   Jordan Tower

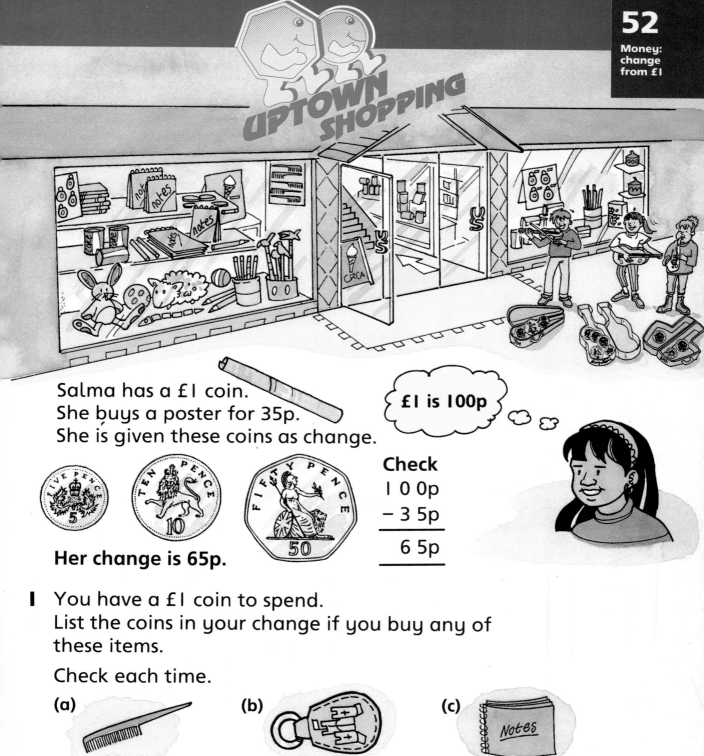

Salma has a £1 coin.
She buys a poster for 35p.
She is given these coins as change.

£1 is 100p

Her change is 65p.

Check
1 0 0p
− 3 5p
6 5p

1 You have a £1 coin to spend.
List the coins in your change if you buy any of
these items.

Check each time.

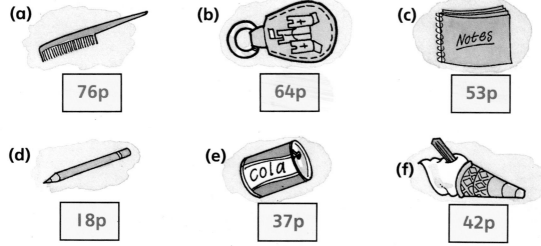

(a) 76p

(b) 64p

(c) 53p

(d) 18p

(e) 37p

(f) 42p

The charity bottle at Pennywise contains 142 pennies.

142p ⟶ 100p + 42p ⟶ £1 + 42p ⟶ **£1·42**

1 In the same way write the amount in each of
these jars in pounds and pence.

(a) 127 pennies (b) 183 pennies (c) 160 pennies (d) 130 pennies

The price tickets at Pennywise are all in pence.

208p ⟶ 200p + 8p ⟶ £2 + 8p ⟶ **£2·08**

2 Change these price tickets to pounds and pence.

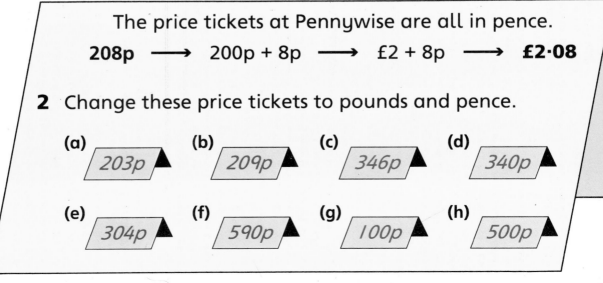

(a) 203p (b) 209p (c) 346p (d) 340p

(e) 304p (f) 590p (g) 100p (h) 500p

1 Write the total amount of money in each busker's case.

2 Which coins would you use to buy each of these items?

List them like this:

scent ⟜ £2·72 ⟶ £1 + £1 + 50p + 20p + 2p

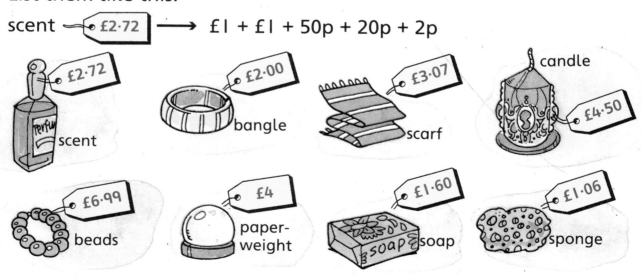

3 ⟜ £2·25 ⟶ £2 + 25p ⟶ 200p + 25p ⟶ 225p ▲

In the same way, show how the price tickets in question 2 would be written in pence.

Fred bought 2 cards.
His bill looked like this.

```
Card     4 7p
Card   + 8 5p
        _____
       1 3 2p  ⟶  £1·32
        _____
```

The total cost of the cards was £1·32.

1 In the same way, copy and complete these bills.

(a)
```
Card     5 6p
Card   + 9 7p
        _____
```

(b)
```
Card     7 4p
Card   + 4 6p
        _____
```

(c)
```
Card     6 2p
Card   + 4 3p
        _____
```

(d)
```
Card     2 9p
Card   + 7 3p
        _____
```

2 Find the total cost of

(a) a pen and a notebook
(b) crayons and glue
(c) ribbon and glue
(d) a pen and a gift tag
(e) glue and a gift tag.

Price list	
pen	89p
notebook	60p
ribbon	55p
glue	52p
gift tag	77p
crayons	68p

```
Pam has £1·25          1 2 5p
She buys a badge for 35p.  − 3 5p
                        _____
She has 90p left.          9 0p
                        _____
```

3 Ben has £1·50. How much has
he left if he buys any of these items?

(a) a pen
(b) a notebook
(c) ribbon and glue
(d) glue and a gift tag

The dog and the rabbit cost

$$
\begin{array}{r}
3\ 7\ 5\text{p} \\
+\ 5\ 2\ 0\text{p} \\
\hline
8\ 9\ 5\text{p} \longrightarrow £8·95 \\
\end{array}
$$

£3·75 is 375 pence
£5·20 is 520 pence

1 Find the total cost of these toys.

(a) dog and panda
(b) cat and dog
(c) rabbit and panda
(d) cat and panda.

2 Find the difference in price between these toys.

(a) panda and cat
(b) dog and rabbit
(c) rabbit and cat
(d) dog and panda.

3 (a) £6·25 + £1·85   (b) £3·43 + £2·08
(c) £8·05 − £5·39   (d) £7 − £5·07
(e) Find the sum of £5·72 and 95p.
(f) £8 minus 37p   (g) 92p plus £6·14

4 You have £4 to spend. How much more do you need to buy these?

(a) game and book
(b) book and calculator
(c) jig-saw and book

GAME £5·74   STORY BOOK £3·36

£4·94

JIG-SAW £2·89

5 Hassan spends exactly £7·83. Which items did he buy?

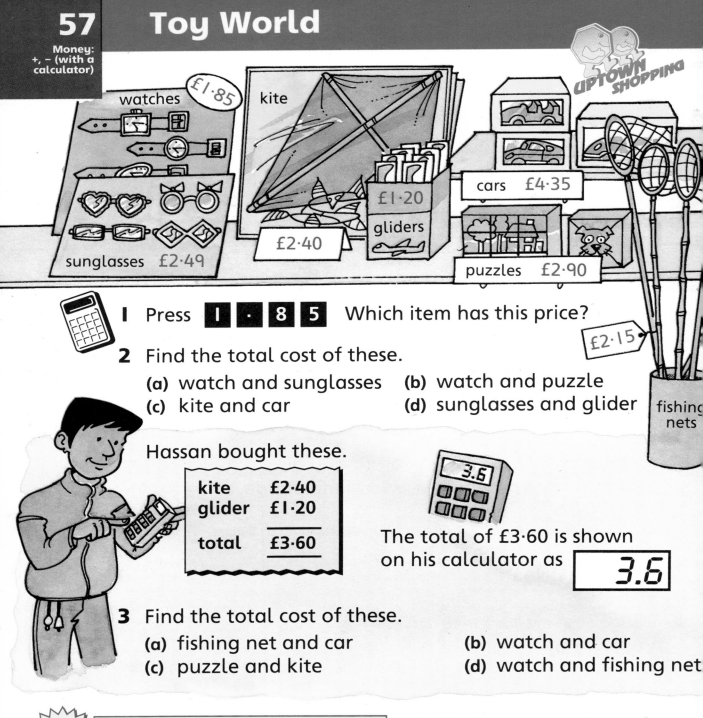

watches £1·85
kite
cars £4·35
gliders £1·20
£2·40
sunglasses £2·49
puzzles £2·90

UPTOWN SHOPPING

£2·15

fishing nets

1 Press **1 · 8 5** Which item has this price?

2 Find the total cost of these.

(a) watch and sunglasses
(b) watch and puzzle
(c) kite and car
(d) sunglasses and glider

Hassan bought these.

kite	£2·40
glider	£1·20
total	**£3·60**

3.6

The total of £3·60 is shown
on his calculator as 3.6

3 Find the total cost of these.

(a) fishing net and car
(b) watch and car
(c) puzzle and kite
(d) watch and fishing net

SALE

£1·25 off these items

Kite £2·40   Puzzle £2·90

Car £4·35   Sunglasses £2·49

4 Find the **new prices** of

(a) a kite
(b) a puzzle
(c) a car
(d) sunglasses.

5 Winston bought one of each item in the sale.
How much did he save altogether?

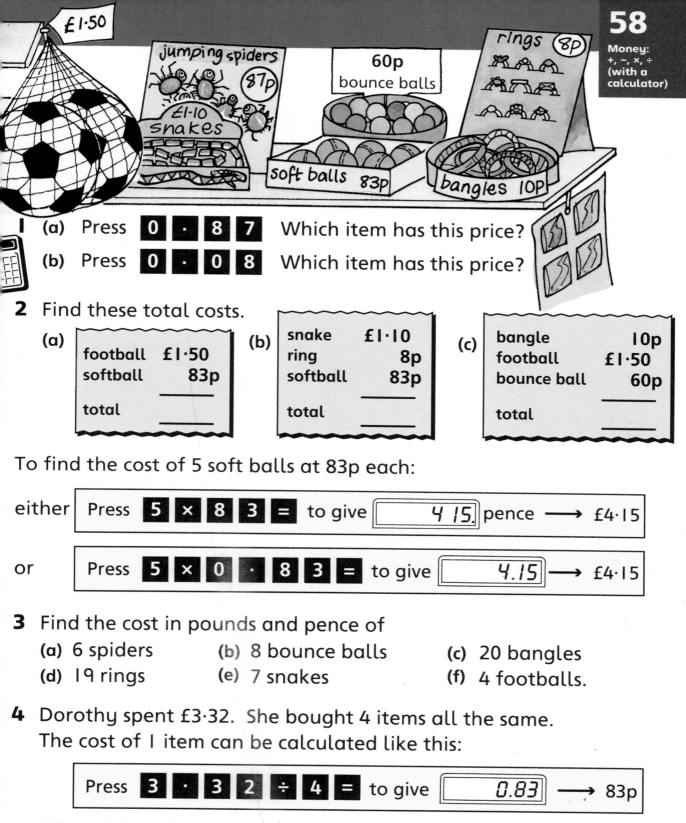

1 (a) Press `0` `.` `8` `7` Which item has this price?

(b) Press `0` `.` `0` `8` Which item has this price?

2 Find these total costs.

(a)
```
football   £1·50
softball      83p
          _____
total
          _____
```

(b)
```
snake    £1·10
ring        8p
softball   83p
          _____
total
          _____
```

(c)
```
bangle        10p
football    £1·50
bounce ball   60p
          _____
total
          _____
```

To find the cost of 5 soft balls at 83p each:

either Press `5` `×` `8` `3` `=` to give ⟦ 4 15. ⟧ pence ⟶ £4·15

or Press `5` `×` `0` `.` `8` `3` `=` to give ⟦ 4.15 ⟧ ⟶ £4·15

3 Find the cost in pounds and pence of
(a) 6 spiders   (b) 8 bounce balls   (c) 20 bangles
(d) 19 rings   (e) 7 snakes   (f) 4 footballs.

4 Dorothy spent £3·32. She bought 4 items all the same.
The cost of 1 item can be calculated like this:

Press `3` `.` `3` `2` `÷` `4` `=` to give ⟦ 0.83 ⟧ ⟶ 83p

What did she buy?

5 Which items did Paul and Mary buy?
(a) Paul spent £7·50 on 5 items all the same.
(b) Mary spent 72p on 9 items all the same.

 Ryan has  £5

He buys a bag for £3·79.
He is given these coins as change.

His change is £1·21.

Check. `5` `−` `3` `·` `7` `9` `=` | `1.21`

1 You have a £5 note to spend.

List the coins in your change if you buy
any of these items.

Check each time.

(a) £2·48

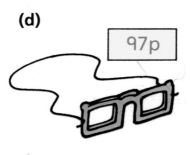

(b)

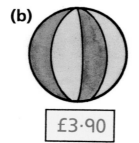

£3·90

(c)

£4·05

(d) 97p

(e)

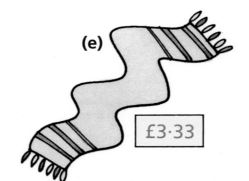

£3·33

(f)

£1·62

Ask your teacher what to do next.

Each twin took half of the apples. | $\frac{1}{2}$ of 12 = 6

They made 2 equal shares. | 12 ÷ 2 = 6

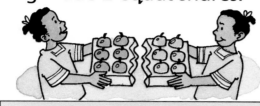

To find one half, divide by 2.

1 Find

(a) $\frac{1}{2}$ of 10   (b) $\frac{1}{2}$ of 8   (c) $\frac{1}{2}$ of 6

(d) $\frac{1}{2}$ of 18   (e) $\frac{1}{2}$ of 2   (f) $\frac{1}{2}$ of 14

Four girls each took one quarter of the pears. | $\frac{1}{4}$ of 12 = 3

They made 4 equal shares. | 12 ÷ 4 = 3

To find one quarter, divide by 4.

2 Find

(a) $\frac{1}{4}$ of 8   (b) $\frac{1}{4}$ of 24   (c) $\frac{1}{4}$ of 28

(d) $\frac{1}{4}$ of 4   (e) $\frac{1}{4}$ of 40   (f) $\frac{1}{4}$ of 32

3 Find

(a) $\frac{1}{2}$ of 16   (b) $\frac{1}{4}$ of 16   (c) $\frac{1}{4}$ of 36

4 There are 20 plums in a box.
One half of them are red and one quarter are yellow.

How many plums are

(a) red   (b) yellow   (c) not red or yellow?

Go to Number Workbook page 23.

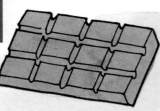

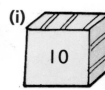
UPTOWN SHOPPING

Three girls each took 1 third of the bar of chocolate.

They made 3 equal shares.

$\frac{1}{3}$ of 12 = 4

12 ÷ 3 = 4

To find one third, divide by 3.

1 Find

(a) $\frac{1}{3}$ of 15   (b) $\frac{1}{3}$ of 21   (c) $\frac{1}{3}$ of 18

(d) $\frac{1}{3}$ of 27   (e) $\frac{1}{3}$ of 3   (f) $\frac{1}{3}$ of 24

To find one fifth, divide by 5.
To find one tenth, divide by 10.

2 Find

(a) $\frac{1}{5}$ of 20   (b) $\frac{1}{5}$ of 35   (c) $\frac{1}{5}$ of 45

(d) $\frac{1}{10}$ of 40   (e) $\frac{1}{10}$ of 70   (f) $\frac{1}{10}$ of 10

(g) $\frac{1}{5}$ of 50   (h) $\frac{1}{10}$ of 60   (i) $\frac{1}{10}$ of 100

3 List the boxes which have the same value.

(a)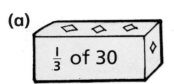
$\frac{1}{3}$ of 30

(b)
30 ÷ 5

(c)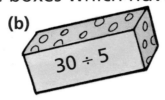
$\frac{1}{5}$ of 30

(d)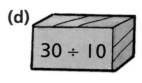
30 ÷ 10

(e)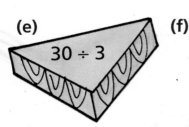
30 ÷ 3

(f)
6

(g)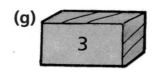
3

(h)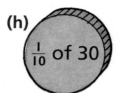
$\frac{1}{10}$ of 30

(i)
10

Ask your teacher what to do next.

Meg has a machine for changing prices in her shop.

1 Meg sets her machine to 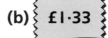 add 5p

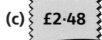 add 5p

13p ◄ IN OUT ► 18p

Find the new prices for

(a) 22p   (b) £1·33   (c) £2·48   (d) £3·07

2 Meg sets her machine to   add £3

Find the new prices for

(a) £18   (b) £1·70   (c) £3·03   (d) £4·99

3 Some items are not selling well.
Meg sets her machine to   subtract 8p

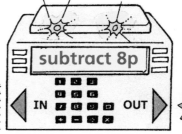

subtract 8p

14p ◄ IN OUT ► 6p

Find the new prices for

(a) 99p   (b) £1·38

(c) £2   (d) £3·06

4 What price change did Meg set for these?

(a)

32p → 39p

£1·22 → £1·29

£3·06 → £3·13

(b)

95p → 89p

£3·42 → £3·36

£7·04 → £6·98

Go to Number Workbook page 29.

Hot and cold

Ask your teacher how to play this game.

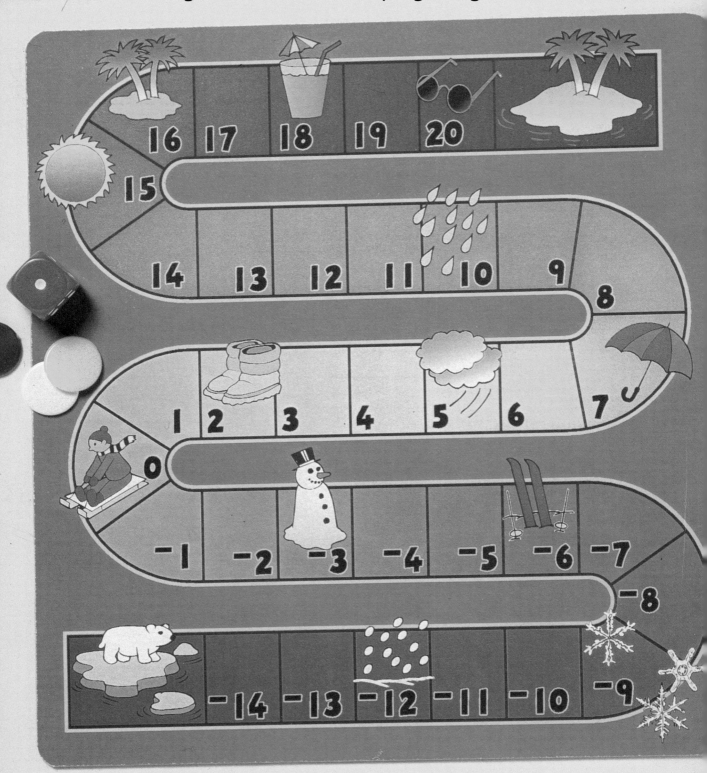

Go to Number Workbook page 30.

1 **(a)** Write this time in words.

(b) Lunch time is 1 hour later. What time will this be?

(c) School started 2 hours earlier. What time was this?

Time words

minutes past
half past
quarter past
quarter to

2 For each of these school clocks, write in words
- the time shown • the time 1 hour later
- the time 2 hours earlier.

(a)

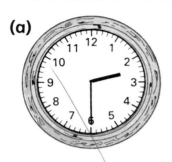

(b)

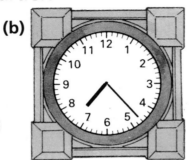

(c)

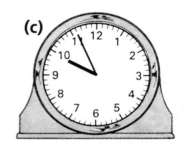

(d)

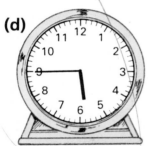

(e)

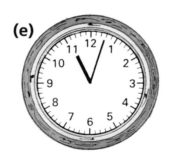

(f)

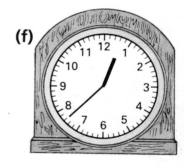

3

Peter's watch shows the time when he left school to go swimming.

(a) Write this time in words.

(b) He came back 2 hours later. What time was this?

4 For each watch, write in words the times
(a) 3 hours later **(b)** 2 hours earlier.

Jan **4:50**

Ali **10:35**

Sue **2:15**

Tom **12:08**

1 The cubs went on a trek. They had to rest often.

(a) At what time did the cubs leave camp?

(b) When did they have their first rest?

(c) For how long had they climbed?

2 How long did the cubs take to complete each part of the trek? Record like this:

from camp to first rest ⟶ I hour

from first rest to second rest ⟶

3 How long did the cubs take

(a) from camp to the top

(b) for the whole trek to the top and back to camp?

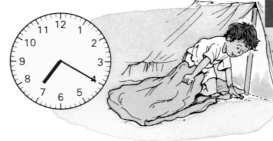

I **(a)** At what time did Ian wake up?   **(b)** When did he get up?

(c) For how many minutes was Ian awake before he got up?

2 When did Ian **(a)** go into the washroom   **(b)** come out?

3 For how long was Ian in the washroom?

4 When did the cubs

(a) start to build the fire   **(b)** light the fire?

5 How long did the cubs take to make the fire?

6 **(a)** When did Ian put eggs in the pan?

(b) When were they ready?

(c) For how many minutes were the eggs boiled?

Clock watching

The time shown is **55 minutes past 2**.

In 5 minutes the time will be 3 o'clock.

The time shown is also **5 minutes to 3**.

1 Write each of these times using **past** and **to**.

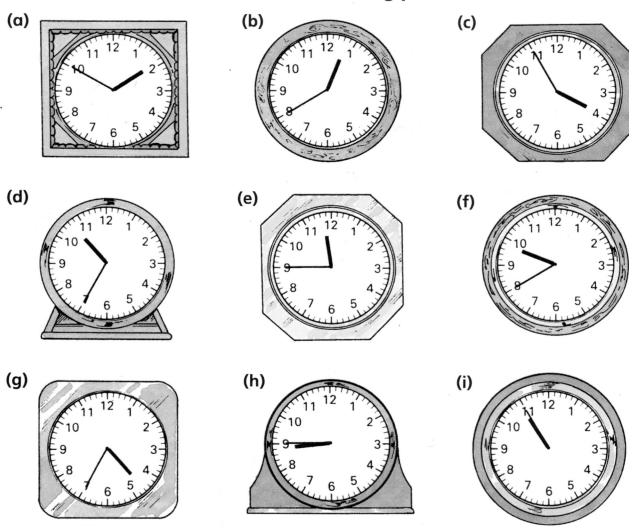

(a) (b) (c)

(d) (e) (f)

(g) (h) (i)

Extension

2 School starts at 9 o'clock each morning.
James catches his bus 25 minutes before school starts.
Write the time, using **past** and **to**, when
James catches his bus.

3 Write the time you leave for school in the morning.

The time shown is **50 minutes past 6**.

In 10 minutes the time will be 7 o'clock.

The time shown is also **10 minutes to 7**.

1 Write each of these times using **past** and **to**.

(a)  11:55

(b)  9:35

(c)  12:50

(d)  4:40

(e)  5:55

(f)  3:45

2 Write each of these as digital times.

(a) 5 minutes to 6   (b) quarter to 4

(c) 25 minutes to 1   (d) 20 minutes to 12

(e) 10 minutes to 10   (f) 15 minutes to 5

3 Write these times in order, starting with the earliest.   Extension

 5:50

 5:35

20 minutes to 6

5 minutes to 6

 5:45

·I FILLEM DENTIST·

I Gail arrived at **5 minutes to 10** for her appointment.
Write the times each of the other children arrived.

Gail

Jack

Lisa

Wes

Peter

Jan

2 Gail, Jack and Lisa were each 10 minutes late.
At what times were their appointments?

3 Wes, Peter and Jan were each 15 minutes early.
At what times were their appointments?

4

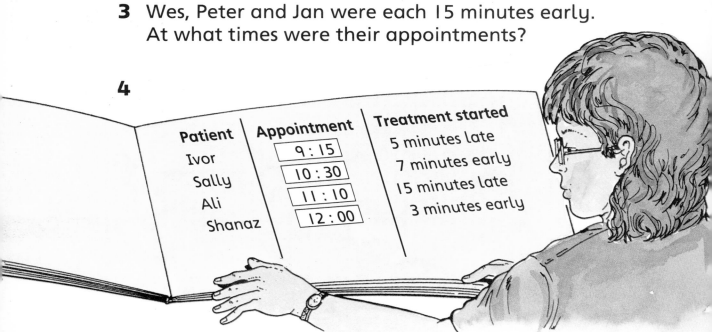

Patient	Appointment	Treatment started
Ivor	9 : 15	5 minutes late
Sally	10 : 30	7 minutes early
Ali	11 : 10	15 minutes late
Shanaz	12 : 00	3 minutes early

Write the time when each patient's treatment started.

Clock repairs

1 Which of Mr Ben's clocks
   (a) show the correct time   (b) are fast   (c) are slow?

2 Which clock is   (a) 5 minutes fast   (b) 10 minutes slow?

3 For each watch, write the correct time.

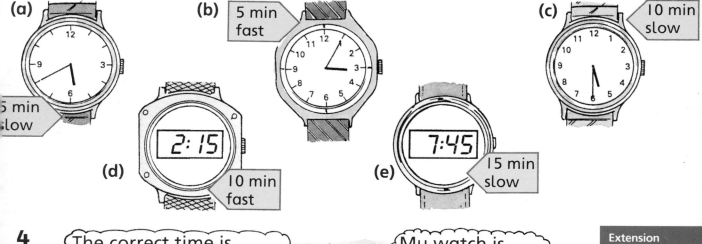

4

The correct time is twenty minutes to nine.

My watch is 10 minutes slow.

Ling

What time is shown on Ling's watch?

The bus left
just before 5 to 9.

The bus arrived at the Park
just after 20 past 10.

Write these times. Use **just before** or **just after**.

1

2

3

4

5

6

1 **(a)** Write in words the time when each programme started.

Cartoon

Filmweek
2:18

In the Wild

Talk show
4:42

Story time

Sportsworld
3:55

(b) Write these daytime programmes in order, starting with the earliest.

2 These programme times have to be changed.

Write each new starting time.

(a)

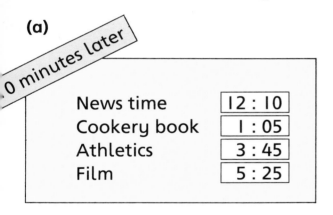

0 minutes later

News time	12 : 10
Cookery book	1 : 05
Athletics	3 : 45
Film	5 : 25

(b)

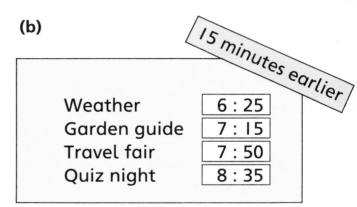

15 minutes earlier

Weather	6 : 25
Garden guide	7 : 15
Travel fair	7 : 50
Quiz night	8 : 35

Problem solving

3 Choose your own order for showing these TV programmes. Give the starting time of each programme. The first one should start at 9:00

| Megapop | 30 minutes | | Superspy | 1 hour |

| Record busters | 15 minutes |

| Laughathon | 60 minutes | | Play scene | 15 minutes |

Go to Measure Workbook page 16.

Drinks

Work as a group.

1 **(a)** Collect **one litre** packs and bottles.
List them.

2 **(a)** Estimate how many children could each have
a glassful from one litre.

(b) Use a litre bottle. Find how many glassfuls
can be poured.

(c) Was your estimate a good one?

3 **(a)** Use your answer from question 2(b).
Copy and complete this table up to 5 litres.

> 1 litre fills about —— glassfuls
> 2 litres fill about ____ glassfuls

(b) About how many glassfuls could be filled
from 10 litres?

Problem solving

(c) How many litres would you need to give
• everyone in the group a glassful
• everyone in the class a glassful
• everyone in the school a glassful?

Work as a group.

1 Use a one litre bottle to find other containers which hold about one litre. List them.

2 Use a one litre bottle. Find a container which holds about:

(a) 2 litres (b) 3 litres (c) 5 litres (d) 10 litres.

3 Ask someone or look up a book to find:

(a) How many litres of petrol the tank of the Headteacher's car can hold.

(b) How many litres of blood an adult has in his or her body.

4 Find how many litres of water a school wash basin can hold.

Problem solving

5 Suzy's family uses 12 litres of milk in a week.

Find out if your family uses more or less than Suzy's family.

Measuring volume

Work as a group.

I Use two I litre bottles.
Find and mark the levels for
(a) I litre   **(b)** $\frac{1}{2}$ litre.

2 Find containers which hold
(a) more than a $\frac{1}{2}$ litre
(b) less than a $\frac{1}{2}$ litre
(c) about a $\frac{1}{2}$ litre.

3 Use a 2 litre plastic bottle.
Mark the bottle to show the level for

(a) $\frac{1}{2}$ litre   **(b)** I litre
(c) $1\frac{1}{2}$ litres   **(d)** 2 litres.

4 Fill the bottle to the 2 litre mark.
Pour out 4 glassfuls. About how much is left?

5 Use your 2 litre bottle to find the
volume of other containers.
Record your results.

6 Here is a chocolate milkshake recipe for two drinks.

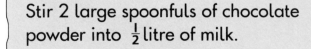

Stir 2 large spoonfuls of chocolate
powder into $\frac{1}{2}$ litre of milk.

Write out this recipe for eight drinks.

Ask your teacher what to do next.

You need 3D shapes like these.

I Match each firework to one of these shape names like this:

Banger box ⟶ cuboid

cylinder	cuboid	cube	cone

pyramid	triangular prism	sphere

2 Look at shapes like these.
Why do you think they are
called **square** pyramids?

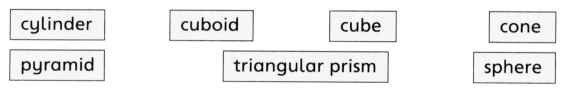

3 What 3D shapes do you see in each of these fireworks?

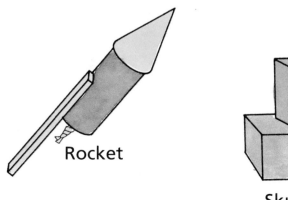

Rocket

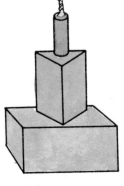

Skyscraper

Blaster

4 Build your own firework using 3D shapes.
Describe the shapes you used.

You need 3D shapes like these.

cuboid

triangular prism

cube

cone

square pyramid

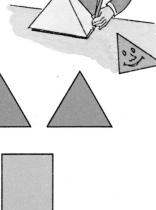
cylinder

1 What is the shape of each red face?

Record like this:

3D shape	Red face
cuboid	rectangle

2 Count and record the total number of faces which each 3D shape has.

Shape	Number of faces
cuboid	6

3 Work as a group.

(a) Take one 3D shape each.
Draw round and cut out each face.

(b) Which 3D shapes have these faces?

Shape 1

Shape 2

1 Which 3D shape is each person talking about?

Anna

Two of its faces are circles.

Steve

All of its faces are the same.

It has one flat face and one curved face.

It has three pairs of rectangles.

Helen

Louise

2 Darren wrote about the faces of a triangular prism:

triangular prism ⟶ 3 rectangles, 2 triangles

What should he write about the faces of these shapes?

cube ⟶

square pyramid ⟶

cuboid ⟶

3 Sue and Andy each sorted their fireworks into two sets by thinking about **faces**.

(a) Write about Sue's sets.

(b) Write about Andy's sets.

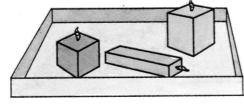

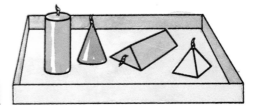

You need 3D shapes like these.

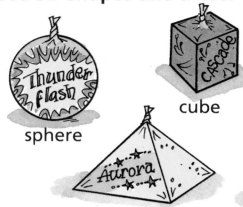

cuboid

sphere

cube

cone

triangular prism

square pyramid

cylinder

A trianglar prism has 6 corners.

1 (a) How many corners does each of the other shapes have?
Record like this: cuboid ⟶ 8 corners

(b) Which shapes have the same number of corners?

A cylinder has 2 curved edges.

2 (a) Which shape has only 1 curved edge?

(b) Which shape has no edges?

3 (a) Make this table.
Count and record the
number of straight
edges for each shape.

Shape	Straight edges
cuboid	
triangular prism	
square pyramid	
cube	

(b) Which shapes have the same number of edges?

4 Which shape is Yasmin
talking about?

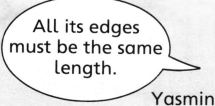

All its edges
must be the same
length.

Yasmin

You need 3D shapes.

1 (a) What is the **same** about these two shapes?
(b) What is **different** about them?

cube   cuboid

2 What is the **same** and what is **different** about these two shapes?

cylinder   cone

3 (a) Which of these shapes have

no curved edges?

List their names.

(b) Which shapes have

no curved edges
and
less than 6 faces?

List their names.

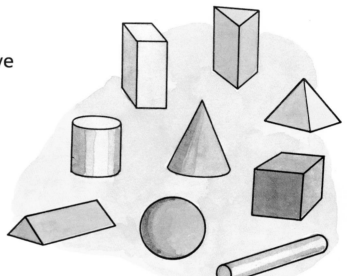

4 Who have made mistakes? Write what they should have said.

It has the same number of faces as corners.

Azif

It has two curved faces.

Debbie

It has more faces than edges.

Hayley

Some of its edges are the same length.

Ben

Ask your teacher what to do next.

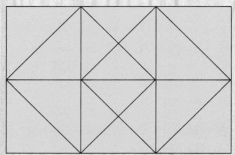

I can see 8 squares.

1 How many squares can **you** see in this drawing?

2 You need a 16-pin nailboard and dotty paper.

The shape on this nailboard is a square.

Make other squares of different sizes.

Draw each square on dotty paper.

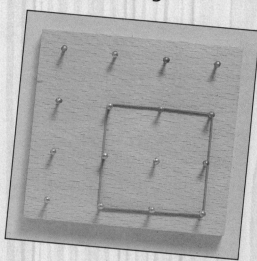

3 You need centimetre squared paper, scissors and glue.

(a) Draw, colour and cut out these five shapes.

> 1 rectangle 5 cm long and 2 cm broad
> 3 rectangles each 5 cm long and 3 cm broad
> 1 square with edge 3 cm

(b) Fit the five shapes together to make a square.
Glue them in place.

(c) Make a puzzle like this for a friend.
Use **four** shapes which will make a square.

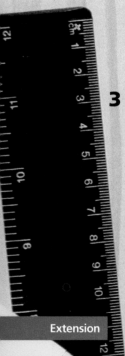

Woodvale school visits Forest Park.

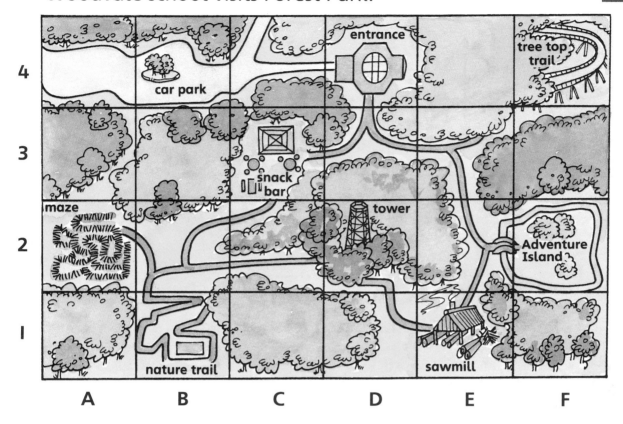

1 What do the children find at
(a) **B1**   (b) **B4**   (c) **C3**   (d) **D4**   (e) **F2** ?

2 At what position is the
(a) maze   (b) sawmill   (c) tower   (d) tree top trail?

3 On Adventure Island,
what is at
(a) **W2**   (b) **X1**   (c) **Y2** ?

4 At what position is the
(a) rope walk
(b) tube slide?

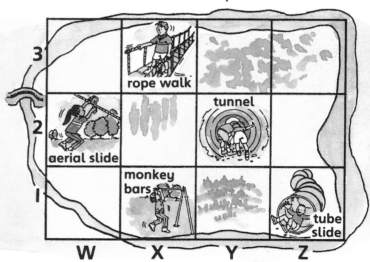

Go to Shape and Handling Data Workbook page 1.

Do Shape and Handling Data Workbook page 2.

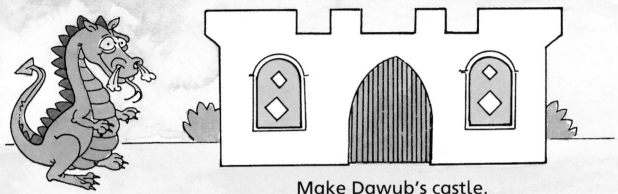

Make Dawub's castle.
It must be symmetrical or Mickey will eat you!

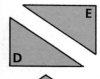

1 You will need a sheet of paper.

(a) Fold the paper in half.

(b) Draw the shaded parts as shown.

(c) Cut away the shaded parts and open out to give Dawub's castle.

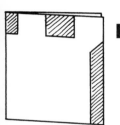

2 (a) Cut out rectangle A from Workbook page 12.

(b) Fold the rectangle in half and draw a curved line as shown.

(c) Cut along the line and open out. This is the gate to Dawub's castle.

(d) Stick the gate on your castle.

3 (a) Cut out rectangles B and C from Workbook page 12.

(b) Put the rectangles together. Fold in half and draw lines as shown.

(c) Cut along the lines and open out. These are the windows of Dawub's castle.

(d) Stick on the windows.

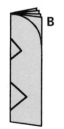

4 (a) Cut out shapes D, E and F from Workbook page 12.

(b) Arrange them to make a symmetrical shape.

(c) Stick this shape above the gate.

You are outside the castle gate.

To enter the castle you must find the password.

You will need some paper and scissors.

I Take 3 sheets of paper. Fold each in half. Fold each in half again.

2 (a) Draw lines on each folded sheet like this.

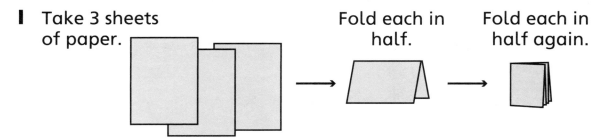

folds   folds   folds

(b) Cut along these lines. Open to find the password.
Each letter has 2 lines of symmetry.

3 Make a shield to hide from Dawub and Mickey.
It may look like one of these.

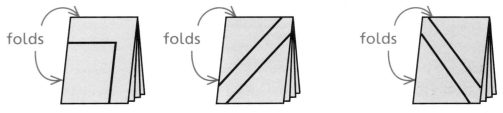

(a) Fold a sheet of paper in quarters.
(b) Draw, cut and open out to make a shield with two lines of symmetry.
(c) Draw a design on your shield.

Go to Shape and Handling Data Workbook page 3.

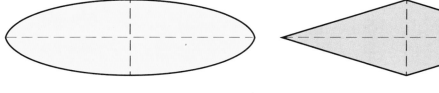

You are at the campsite.

1 What do you see if you look   **(a)** north   **(b)** south?

2 In which direction is   **(a)** the castle   **(b)** the harbour?

3 Face north.

In which direction are you facing if you turn

(a) 1 right angle clockwise   **(b)** 2 right angles clockwise

(c) 3 right angles anticlockwise   **(d)** 4 right angles clockwise?

4 Face east.

What do you see if you turn

(a) 1 right angle anticlockwise   **(b)** 2 right angles clockwise

(c) 3 right angles clockwise   **(d)** 4 right angles clockwise?

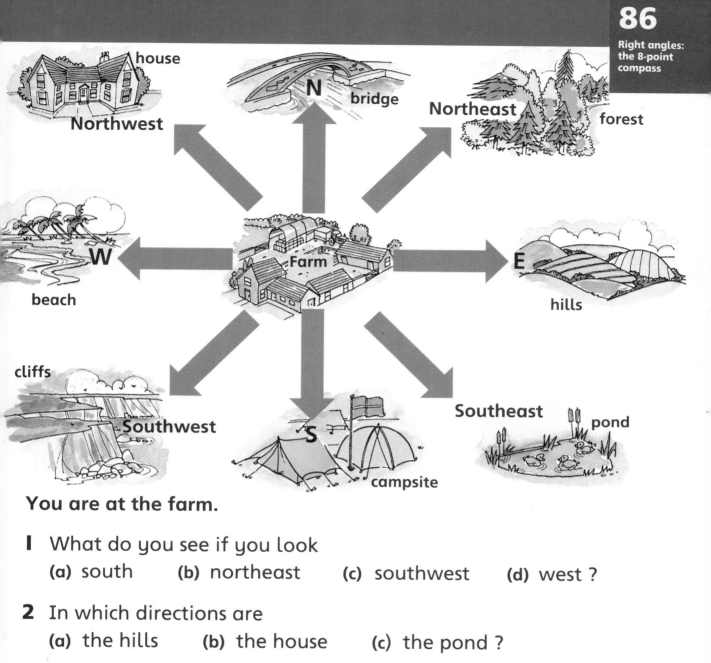

You are at the farm.

1 What do you see if you look
   (a) south   **(b)** northeast   **(c)** southwest   **(d)** west ?

2 In which directions are
   (a) the hills   **(b)** the house   **(c)** the pond ?

3 Face north.
   In which direction are you facing if you turn
   (a) 1 right angle clockwise   **(b)** 2 right angles anticlockwise
   (c) 3 right angles anticlockwise   **(d)** 4 right angles clockwise?

4 Face south.
   In which direction are you facing if you turn
   (a) 2 right angles clockwise   **(b)** 4 right angles?

Ask your teacher what to do next.

Woodvale School *Fundraising Events*
Saturday 14th June to Saturday 21st June

When	What	Where
Sat 14th	school fayre	school playground
Sun 15th	sponsored walk	Forest Park
Mon 16th	readathon	Woodvale library
Tue 17th	quiz	village hall
Wed 18th	concert	school hall
Thu 19th	concert	school hall
Fri 20th	concert	school hall
Sat 21st	concert	school hall

1 (a) In which month were the fundraising events held?

  (b) On which day was • the sponsored walk? • the quiz?

  (c) Which events were **not** held at Woodvale School?

  (d) Which events were held • indoors? • outdoors?

Woodvale School Fayre

Each event only 50p per person

Number of people at each event

Starting times	Soak the teacher	Bouncy castle	Custard pie fight	Crazy games	Paint your parent
9:00	10	4	8	8	9
9:30	15	8	10	12	5
10:00	25	20	30	24	16
10:30	22	16	22	20	10

2 (a) How many people took part in the 10:00 Crazy games?

  (b) How many people took part in the 9:30 events?

  (c) What was the most popular event?

  (d) How much money was made from the 10:30 events?

Extension

  (e) It was hoped that all the events would raise £130.
Did the school meet this target?

Notice board				
Woodvale School concert				
	Wed	Thu	Fri	Sat
Choir	✓		✓	✓
Play	✓	✓	✓	✓
Dancers		✓		✓
Display	✓	✓	✓	
Band	✓		✓	✓
Magician	✓	✓		
Singer		✓	✓	✓

3 **(a)** Which item is performed **every** evening?

(b) On what evenings do the dancers perform?

(c) On what evening do the choir **and** the dancers perform?

(d) What items are **not** performed on Saturday?

(e) How many items are performed each evening?

Woodvale School
Concert Programme
7:30 – Choir
7:50 – Display
8:15 – Singer
8:25 – Band
8:40 – Interval
8:55 – Play

4 **(a)** Which item started at
   • quarter past eight?   • five minutes to nine?

(b) Which item finished just before
   • twenty-five minutes past eight?   • ten minutes to eight?

(c) Which item was being performed at 8 o'clock?

(d) How long was the interval?

(e) The concert finished at half past nine.
   How long was the play?

5 Make up a programme of your own giving starting
times and items.

Go to Shape and Handling Data Workbook page 15.

Vera's Video Shop

Do Shape and Handling data Workbook page 17.

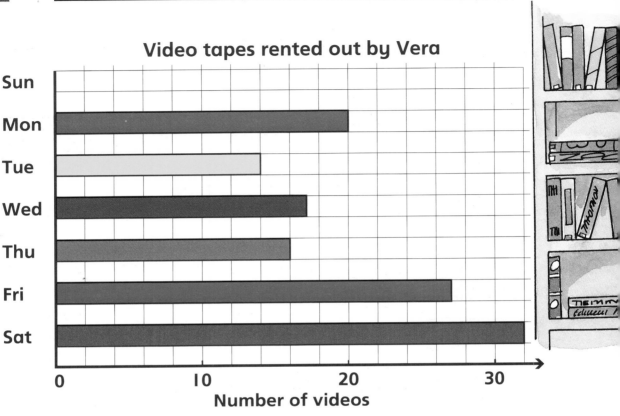

Video tapes rented out by Vera

Number of videos

1 How many videos were rented on

(a) Monday   (b) Tuesday   (c) Wednesday   (d) Friday?

2 How many videos were rented on Sunday? Explain this.

3 How many more videos were rented on Saturday than on Tuesday?

4 On which day were twice as many videos rented as on Thursday?

5 Use this information to draw a graph.

Videos rented last week

Sun	Mon	Tue	Wed	Thu	Fri	Sat
0	14	17	22	15	29	30

Each represents **5** buses.

 represents **between 10 and 15** buses.

Bus trips to Forest Park

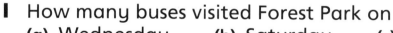

Sun	
Mon	
Tue	
Wed	
Thu	
Fri	
Sat	

1 How many buses visited Forest Park on
   (a) Wednesday **(b)** Saturday **(c)** Tuesday?

2 On which day did **between** 25 and 30
   buses visit?

3 On which days did **fewer** than 10 buses visit?

4 On which days did **about the same**
   number of buses visit Forest Park?

5 Why do you think Saturday and Sunday
   were busier days?

Go to Shape and Handling data Workbook page 18.

For each event write: **certain, very likely, likely, unlikely, very unlikely** or **impossible.**

1 I will become Prime Minister.

2 I will be on holiday from school on 2nd August.

3 I will grow to be 40 metres tall.

4 I will watch TV at school this week.

5 I will find a £1 coin this week.

6 I will eat some vegetables tomorrow.

1 Which is **more likely?**

A

B

2 Which is **less likely?**

C

D

3

| January | June | September |

Which of these months is
(a) most likely
(b) least likely
to be on this page of the calendar?

4 Mel picks a lollipop from
the box without looking.

Which colour is he
(a) most likely to pick
(b) least likely to pick?

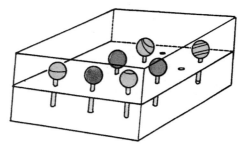

5 Draw pictures of two events and ask your friend
which one is more likely.

Pick a ball.
If it's red Suzi wins.
If it's blue Tom wins.

Look at the colour of the balls.

1 Is this game fair? Explain your answer.

2 (a) Is this game fair? Explain your answer.
(b) Who has a better chance of winning?
(c) How would you make the game fair?

> If there are the same number of
> red and blue balls, Tom and Suzi
> have an **equal** or **even** chance of winning.

3 (a) For each of these games, is Suzi's chance of winning
more than even, even or **less than even**?

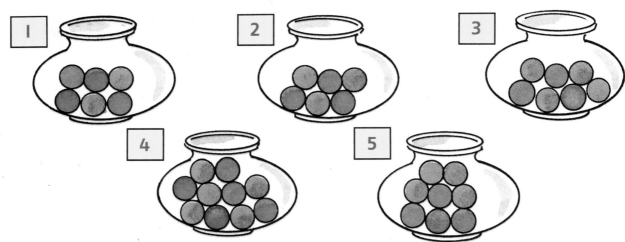

| 1 | 2 | 3 |
| 4 | 5 |

(b) For the unfair games, say how you would give
each person an even chance of winning.

Go to Shape and Handling Data Workbook page 22.